The Pedagogy of Physical Science

Contemporary Trends and Issues in Science Education

VOLUME 38

SCOPE

The book series Contemporary Trends and Issues in Science Education provides a forum for innovative trends and issues connected to science education. Scholarship that focuses on advancing new visions, understanding, and is at the forefront of the field is found in this series. Accordingly, authoritative works based on empirical research and writings from disciplines external to science education, including historical, philosophical, psychological and sociological traditions, are represented here.

For other titles published in this series, go to
www.springer.com/series/6512

David Heywood · Joan Parker

The Pedagogy of Physical Science

Springer

David Heywood
Manchester Metropolitan University
Institute of Education
Manchester
Didsbury
United Kingdom
d.heywood@mmu.ac.uk

Joan Parker
Manchester Metropolitan University
Institute of Education
Manchester
Didsbury
United Kingdom
j.parker@mmu.ac.uk

ISBN 978-1-4020-5270-5 e-ISBN 978-1-4020-5271-2
DOI 10.1007/978-1-4020-5271-2
Springer Dordrecht Heidelberg London New York

Library of Congress Control Number: 2009942129

Printed on acid-free paper

Springer is part of Springer Science+Business Media (www.springer.com)

Acknowledgements

The inception of this project derived from a professional concern about effective teaching to support meaningful learning in science. It is an attempt to develop insight into effective pedagogy from the perspective of the learner. The book draws on researching pre-service and practicing teachers' learning of science on teaching programmes at Manchester Metropolitan University over the last decade. We would like to thank all those students and teachers who gave freely of their time and participated enthusiastically in such a way that allowed us to gain insights that would have otherwise been impossible.

We offer thanks to our colleagues in the science education department past and present for their engagement in lively discussion about the ideas we were struggling with. In no particular order we are indebted to Mark Rowlands for his carefully considered insights, Frank Gibson for his intellectually challenging questioning, Alan Goodwin for his limitless enthusiasm, Denis Burns for his inspiration and Gill Peet for never tiring of being interested in our work.

A special thanks to all the staff of the Education and Social Science Research Institute for their continued interest and faith in our research work. In particular we appreciate the support of Harry Torrance, Liz Jones and Maggie McClure and, especially Tony Brown, who encouraged us from the outset and gave us the belief that we were able to undertake such a daunting project.

We are grateful to the editors of the *British Education Research Journal*, the *Cambridge Journal of Education and* the *International Journal of Science Education* for permission to draw on previously published articles in the completion of this work.

The script has been painstakingly revised on numerous occasions as a result of valuable critical comment from those who gave freely of their time to help with the initial and final drafting of the text. Sincere thanks to Ann Heywood and Rob Heywood for their help in this.

Finally, a special mention goes to our families, Ann and Rob, Phil, Lucy and Kate for their patience, support and encouragement.

About the Authors

Dave Heywood is Reader in Education and **Joan Parker** Senior Lecturer in Education at the Manchester Metropolitan University Institute of Education, England. Their research interests are focused on developing teacher subject and pedagogic knowledge. They work collaboratively with colleagues to research how pre-service and practicing teachers develop science subject and pedagogical knowledge in order to enhance higher education taught provision in Initial Teacher Education (ITE) and Continuing Professional Development (CPD) programmes. The research undertaken provides evidence that by engaging in a metacognitive approach to learning, in what are problematic subject areas for many science trainees and teachers, there resides the opportunity to foster not only understanding of scientific concepts but also pedagogical insight into the learning of them. Their current research interests concern the development of this approach and exploration of its application in the classroom practice of teachers. They have recently been involved in working with the Manchester Museum of Science and Industry (MOSI) education officers and school teachers in promoting out of school learning to develop pupils' enthusiasm and confidence in science, technology, engineering and mathematics (STEM). They are passionately committed to focusing research on practice to inform future programme provision for both ITE and CPD. They have published internationally and presented at conference both nationally and internationally.

Contents

Chapter 1
Introduction

Pedagogy, the principles and practices of teaching, is a central concern in science education and has formed the focus of much educational research over the last 2 decades. This book focuses on the process of how subject and pedagogic knowledge emerge through teachers' learning in science. It draws on a substantive body of empirical research, collated over the past decade, focusing on conceptual domains that are known to be difficult for learners including forces, electricity, light and basic astronomy. The findings are derived from analysing pre-service and practicing teachers' responses to engaging with difficult ideas when learning science in higher education settings. In an effort to address the questions regarding problematic science concepts in their own learning, the teachers in the studies we report here are themselves afforded an opportunity to focus on the nature of the concepts being explored and the manner in which an understanding of them might be developed; they are, therefore, referred to as learners or students of science throughout.

Despite recent relative success in achievement as measured by knowledge acquisition, there is an increasing concern with about the problem of pupils failing to see meaning in the ideas they encounter in their science learning. The issue remains a significant one, not least in regard to the lack of interest in the subject pupils exhibit in the subject as evidenced in the declining uptake of the sciences in higher education. While the factors that impact on this are multifaceted, the importance of teachers developing sufficient confidence to teach science creatively in order to engage and enthuse pupils' learning of science is likely to be a significant contributing factor. The breadth and depth of curriculum content in science has placed a considerable demand on teachers' subject knowledge. Subsequently, this has had significant influence on both initial and in-service teacher education. The tensions in regard to supporting teachers in developing their subject knowledge so that they feel confident in teaching science in interesting, challenging and creative ways are difficult to reconcile. This is particularly the case in respect of pre-service teachers who are non-specialists in the subject.

The conceptual demand of science places the teachers' subject knowledge at the heart of developing confident and competent practitioners. Teachers require not only a sound and secure base of subject knowledge, but also the ability to implement

D. Heywood and J. Parker, *The Pedagogy of Physical Science*, Contemporary Trends and Issues in Science Education, vol. 38, DOI 10.1007/978-1-4020-5271-2_1,

a range of teaching and learning strategies to develop appropriate explanations to support learning. This entails a synthesis of both subject and pedagogy and there is a need for explicit exemplification of what such pedagogic knowledge might be within specific science domains. This book addresses some of the implications arising from this.

In response to increasing the accountability of educational institutions by government agencies, there is a danger of over-emphasis on the assessment of student knowledge of facts. In teacher education, this has been evidenced by the emergence of an auditing and testing culture. This can be construed as portraying science education as a process of information transfer and recall, as opposed to one of developing ideas and explanations. One way of countering such an unproductive view of science is to ensure that during their training, students are provided with teaching and learning experiences that are designed to challenge this view of teaching. This could encourage them to consider the nature of both their own learning and that of children through carefully reviewing direct learning experiences.

Although developing personal subject knowledge for teachers is often framed within a deficit model in which initial teacher education attempts to support students in addressing areas of weakness, we propose that the very act of identifying and addressing problematic science concepts in their own learning affords an opportunity for students to focus on the nature of the concepts being explored and how understanding of them might be enhanced. This constitutes a productive way of turning a deficit model of teachers' subject knowledge into a positive experience with considerable potential for the development of pedagogy. It is a central theme developed throughout and is based on purposefully presenting the problematising of the subject as a positive condition of professional being through which insights into pedagogy emerge that would otherwise remain latent. We contend that this approach is more likely to lead to both conceptual and pedagogic change. The former is recognised as an integral and necessary element of learning science because it is often required to make sense of what initially appear to be counterintuitive explanations of the world. The latter, whilst clearly a core professional concern and valued goal in science education is not as well-articulated. It concerns the professional issue of interpreting and constructing coherent causal explanation for phenomena that serve to provide a convincing account that both persuades and engages learners because it makes learning meaningful to them. Our work here is an attempt to inform contemporary debate on this issue. It argues that the deliberate presentation of science learning as problematic (for both teacher and pupil) is both a necessary condition and a positive conceptualisation of what it is to learn science and can be used productively in promoting not only knowledge and understanding of science, but also valuable pedagogic knowledge of teaching and learning.

The research reported here is based on findings from empirical studies undertaken at Manchester Metropolitan University in England. In order to achieve Qualified Teacher Status (QTS), students on pre-service higher education programmes must demonstrate that they have the required subject and pedagogic knowledge in science to teach effectively. The stipulated requirements for pre-service teacher standards are outlined by the Training and Development Agency (TDA 2007) for schools which reflect the demands of the school science curriculum. The school curriculum referred to is the English National Curriculum for schools (DfEE/QCA 1999) which has four Key Stages (KS 1–4). These are divided into year groups from year 1 to year 11 (Y1–Y11). The first two Key Stages (KS1 and KS2) are undertaken in primary schools (Y1–Y6) with Key Stages 3 and 4 (Y7–Y11) being completed in secondary school.

In order for the methodology to be coherent with the pedagogical approach adopted, it was felt necessary to explicitly acknowledge and share with students that teaching sessions were research-focused. The principal objective of the empirical studies throughout was the synthesis of research and teaching for the purposes of developing insight into the learning process. This was intended to explore a range of issues including the identification of sequences in cognition and to address the extent and limitations to which this could be paralleled with a sequence in pedagogy. A key principle in the methodology concerned securing the student perspective during the tutor and peer group discourses within taught sessions at university. The epistemological basis for this approach is different from that adopted in pre- and post-teaching evaluations of student understanding. The process attempts to capture a 'dynamic'; it places considerable (metacognitive) demand on the learner and requires them to identify and articulate significance in their own learning. To this end, student written accounts, annotated drawings, session notes and recorded discussions were collated and analysed to identify patterns that could provide insight into those elements that they found useful in developing meaningful interpretations of abstract ideas. A key element of tracking learning involved students in keeping a reflective journal to document their engagement with initial thinking about, and subsequent engagement with, ideas encountered in the teaching sessions to consider how the experience impacted on their perceptions of pedagogy. The process generated significant insights into factors influencing the emergence of pedagogy. The qualitative data that comprised the basis of the analysis was drawn from interviews, discourses, reflective journals and summative assignment writings. Written journal entries were a primary data source and in some cases, where meaning was ambiguous, students' ideas were discussed further at interview. This data was subsequently scrutinised through reviewing summative assignment tasks to determine the extent of coherence in reasoning.

It is important to recognise that the notion of problematising science subject knowledge requires analysis that is necessarily interpretative. Attempting to document the process of change in students' perceptions through a qualitative approach within an interpretivist paradigm derived from accounts of their own learning is applicable to both the researcher and the learner. Tutors initially determined what (through anticipation in planning) would constitute 'critical incidents' in learning, such as typical cases where cognitive conflict was likely to ensue as students explored the various phenomena through practical investigation. The subsequent data analysis process had a significant impact on programme development and provision. The presenting of data as narrative from students' responses is a feature of all the studies cited. Whilst earlier studies, when working with large groups of students focused on finding patterns to categorise key ideas, the emphasis in subsequent research moved increasingly towards the presentation of data as narrative, although categories of key ideas has remained a feature. Through the act of analysing their own learning in this way, students were able to identify problematic aspects for learners in developing specific scientific ideas as well as developing insights into a range of general pedagogic implications. The physical science studies show that pre-service teachers are able to generate important insights into the nature of scientific ideas and the learning of them through such a process.

Critical features of developing a metacognitive approach include the need to create a learning environment of trust and security between tutors and students such that learners are confident in sharing perspectives. It requires time and opportunity to nurture the socio-cultural environment of learning in which knowledge is problematised. In some ways such an approach can be said to militate against the current teacher training trend towards diminishing course contact time, development of distance learning materials and ICT dependency. Factors such as personal involvement in learning and ownership of learning were important in creating a positive environment in which students were not afraid to discuss their thinking. Discussion was identified as a central feature of the learning process. There were important pedagogical insights into the teaching and learning of specific subject matter as explicitly identified by the students across the studies. This is exemplified and detailed in Chapter 7.

Through introspection and collective discussion of perspectives, students are, in effect, 'auditing' specific science subject domains such that they become aware of typical misconceptions and inherent difficulties in developing understanding of them. These features, although specific to particular subject matter, can be used to alert the teacher to likely problems in other areas. Having, for instance, recognized that there is a need to differentiate spin and orbit in translating written information about day and night and seasons, teachers can be alerted to audit other subject areas for similar potential language problems such as current flow and energy transfer in understanding the lighting of a bulb in a simple circuit.

In teacher education, research into teacher learning during university teaching sessions offers significant potential for further developing insight into pedagogy because it provides opportunity for a unique synthesis in which the students are reconciling experience as both teacher and learner. The process of problematising subject knowledge through direct experience of learning in areas of science that are known to be difficult constitutes the most productive way of realising this potential, turning a deficit model of teacher subject knowledge into a positive learning experience. The book addresses these issues in the following way:

Chapter 2 presents a review of conceptual change literature. This has had a pervasive influence on science education research over the last 2 decades, informing the direction of focus for studies that have generated significant insight into the problem of how to promote conceptual change in learners across a range of domains in science learning. We discuss various attempts to develop models of conceptual change and the theoretical rationale that underpins these and provide accounts from student learning about forces to contextualise these debates in relation to subject and pedagogic knowledge.

Chapter 3 examines pedagogic approaches to the learning and teaching of electricity through analogy describing how students apply a range of analogies to developing understanding of simple electric circuits. The evidence indicates that all analogies break down under critical scrutiny. Deriving from this we argue that the pedagogic task should shift emphasis from the search for the 'holy grail' of analogies that attempts to comprehensively address learners' concerns at understanding the intangible. Instead, we suggest that pedagogy should explicitly recognise and focus on the generative power of analogies in promoting thinking when they break down under application to increasingly difficult scenarios.

Chapter 4 explores the role of cognitive conflict in promoting learning. It draws on evidence from student learning about light and shadow and presents an account of how varying degrees of cognitive conflict can be used productively to promote significant insight into pedagogic issues in teaching about abstract phenomenon.

Chapter 5 questions some of the traditional assumptions implicit and explicit in science learning where language is often conceptualised as a tool to be applied to making sense of difficult ideas. In challenging this notion we offer an account of how language works as a structure and provide exemplification of the ways in which understanding is produced by rather than reflected in language. The emphasis is on interpretation and the possibilities and constraints that language imposes on us, questioning both the epistemology and ontology of how science is represented in the curriculum.

Chapter 6 is concerned with professional learning developing an understanding of how the process of metacognition can be used to inform pedagogic insight into teaching about basic astronomy. It documents students engaging with abstract ideas using a range of strategies including modelling and juxtaposing between two- and three-dimensional representations in order to identify generic pedagogic implications in learning.

Chapter 7 evaluates the impact of students reconciling experience as both teacher and learner in confronting difficult ideas in terms of developing insight into the professional dimension of planning for the appropriate representation of science in the curriculum. It explores the application of what we term a Subject Matter Learning Audit approach to student learning and illustrates how this impacts on emerging pedagogical insights including the identification of key ideas and associated teaching points for the effective presentation of difficult and abstract ideas to primary-aged children.

Chapter 2
Conceptual Change and Learning About Forces

This chapter outlines some aspects of the historical evolution of research into science learning and examines some of the complexities of the conceptual change process in learning about forces. Section 2.1 discusses the influence of cognitive psychology, in particular Piaget's work on the individual construction of meaning, and how this relates to classical models of conceptual change. The discussion includes a review of conceptual change models concerned with developing knowledge and understanding of learners' conceptions in science and explores some of the more recent criticisms of such approaches. This section concludes with a brief examination of socio-cultural and social constructivist perspectives. Section 2.2 provides an empirically based account of conceptual change in action detailing primary teachers' learning about forces. This part of the discussion explores the generation of an emergent pedagogy as teachers analyse the dynamics of their own learning.

2.1 The Challenge of Learning About Forces and Motion

> The physics of motion provides one of the clearest examples of the counter-intuitive and unexpected nature of science. (Wolpert 1992:2)

Forces and motion constitute one of the most challenging areas of study in science education; a multitude of research attests to its difficulty (for example, see Brown and Clement 1987; Kruger et al. 1990; Driver et al. 1994; Tao and Gunstone 2000). The teaching of force and motion involves the development of abstract concepts that are likely to be at odds with learners' everyday life experiences. Intuition drawn from such experience supports the idea that a force is an 'entity' that is 'given to' or 'resides within' an object, causes it to move and is 'used up' in motion. In pushing a toy car across a carpet, for example, it would seem reasonable to assume that the force has been given to the car by the initial 'push' causing the car to move fast at first, gradually slowing down and stopping as the push is used up. This contrasts sharply with Newtonian physics of motion, and, in teaching about forces, the teacher constantly faces the challenge of moving learners from this intuitive

D. Heywood and J. Parker, *The Pedagogy of Physical Science*, Contemporary Trends and Issues in Science Education, vol. 38, DOI 10.1007/978-1-4020-5271-2_2,

view of force residing within objects towards the scientific notion of forces acting on objects. The difficulty in making such a transition is that learners' conceptions of force as a consumable entity are perpetuated and continually reaffirmed through everyday experiences of moving objects. For instance, in pushing someone on a swing or kicking a football, the harder one pushes, the longer it takes for the swing to stop or the stronger the kick, the further the ball moves. Consequently, if a scientific explanation based on Newtonian physics is to be made intelligible and useful for learners, there needs to be a radical shift in intuitive explanatory frameworks. This is likely to require significant restructuring in thinking and, as we shall see in this chapter, as constructs such as forces acting *on* objects and weight *as* a force constitute abstract and deeply counterintuitive ideas that create conceptual difficulties for learners grappling with Newtonian mechanics at a qualitative level of thinking. They form a key element of this discussion as we explore pre-service and primary teachers' learning in action as they experience the process of conceptual change in this area.

Force and motion forms part of teaching and learning in the primary school in many countries where it is specified in national curricula, curriculum guidelines or encountered within an inquiry-based approach to science education. It is important, therefore, to consider the knowledge and understanding that teachers will need to possess with regard to forces in order to structure learning for children. In order to deliver the current English primary science curriculum effectively, teachers will need to possess an understanding of balanced and unbalanced forces and their effect on motion (DfEE/QCA 1999). They will need to understand that when an object is stationary or moving at a constant speed in a straight line, the forces acting on it are balanced, whereas unbalanced forces result in a change in motion, shape or direction. As primary teachers and pre-service teachers often possess naïve conceptions in physics such as those illustrated above (Kruger et al. 1990; Parker and Heywood 2000; Heywood and Parker 2001), a considerable degree of conceptual shift is needed in order to meet curriculum demands, and this presents a major challenge for primary teacher education. The process of this changing shift in understanding is explored in our account of primary teachers' and pre-service teachers' learning in this domain as they investigate a typical range of educational contexts involving force and motion. We look at how intuitive ideas can be influenced in teaching and learning and explore the difficulties and challenges teachers encounter in constructing a qualitative understanding of force and motion within common classroom activities. First, we will consider something of the nature of the conceptual change process.

2.2 Conceptual Change: A Brief Historical Perspective

In the study of cognition, concepts are abstract or psychological constructs that represent 'ideas' or 'notions' that a learner uses in reasoning and thinking. They constitute the general tools of enquiry used in making sense of the world.

In science, learners' existing concepts are known to have a profound influence on how phenomena are interpreted, and learners draw on these concepts in making predictions and explaining what they see and experience in the world (Ausubel 1968). The recognition of this underpins the predominance of a broadly constructivist paradigm in science education in which science learning is perceived as an active process of personal construction whereby knowledge, understanding and experience of the world are integrated with new ideas and experiences.

Research has contributed much to our knowledge and understanding of science learning in the last three decades (Fensham 2001; Duit and Treagust 2003; Jenkins 2004). It has been concerned with the search for a theoretical model of learning that better informs instructional strategies to support the development of understanding. The focus on the nature of concepts, their internal organisation and what influences changes in them has provided insight into a variety of perspectives on learning from which has emerged a range of practices (see, for instance, DiSessa and Sherrin 1998; Tao and Gunstone 2000; Limón and Mason 2002). The following discussion highlights some significant developments in this field. For a more comprehensive consideration of related research see Abell and Lederman (2007).

2.2.1 *The Influence of Piaget*

The roots of conceptual change research derive from the work of Swiss philosopher, genetic epistemologist and developmental psychologist Jean Piaget (1896–1980). His seminal theory of cognitive development presents an interactive view of learning whereby learners make sense of the world through cognitive schemes that evolve as children interact with objects in their environment; it has had a profound effect in educational thinking (Piaget 1964). Although Piaget clearly considered the social dimension of children's interaction in the world as an important feature of learning, he was essentially concerned with cognitive development as a result of maturation, rather than learning instruction or social interaction. In this sense, he could be considered to be a radical constructivist.

Piaget considered that changes in the organisation of cognitive schemes occur as a result of interactions in the world; a process he called adaptation that involves two associated processes (assimilation and accommodation). Assimilation results from an individual interpreting sensory information from the environment that subsequently is incorporated into the individual's mental cognitive schemata. Accommodation is the process by which the cognitive structure is adapted to make sense of the information. Both processes occur when sensory information is received. In addition, his work with Inhelder in 1956 on content-independent logical structures as stages in the development of logical thinking had a pervasive influence on emergent approaches to primary education. For instance, during the 1960s, Piaget's work informed a major review of English primary education (Plowden

1967) that extolled the virtues of experiential learning. In particular, the Concrete Operational Stage of Development located between the ages of 2–12 years, which involves the co-ordination of conceptual schemes such as conservation, classification and seriation, was to influence primary curriculum design. At this stage of development, it was proposed that children are not capable of performing operations at the purely symbolic level, and this thinking had a powerful impact on educationalists' views of what was possible for learners at different stages of maturation. Although questions have been raised about the Piagetian stages at which children can perform tasks and the influence of the context in which they are carried out (for example, Donaldson 1978), his work subsequently informed a classical model of conceptual change that was to become pivotal within the field of conceptual change research in science education.

2.2.2 *The 'Classical' Model of Conceptual Change*

Developed in 1982 by Posner et al., and later refined and developed (Hewson 1981, 1982, 1996; Hewson and Hewson 1984, 1988; 1992; Strike and Posner 1992), the model forms the basis of conceptual change tradition. Drawing on ideas from the domains of developmental psychology and the philosophy of science, it synthesises Piaget's views on assimilation and accommodation with the ideas of Kuhn (1970), Lakatos (1972) and Toulmin (1972) in relation to the historical development of scientific knowledge and understanding. It also resonates with Ausubel's contention that an individual's existing knowledge and understanding is the most significant influence in learning (Ausubel 1968). The central tenet of the conceptual change model is that of the learner becoming *dissatisfied* with an existing conception as a prerequisite to initiating radical (or revolutionary) conceptual change, a process which explicitly parallels that of 'paradigm shift' in science.

If conditions of learning precipitate dissatisfaction and an *intelligible*, *plausible* and/or *fruitful* replacement conception is made available to the learner, then conceptual change will ensue. This model focuses on the accommodation of the replacement conception and the conditions under which this will occur. The new conception must be sensible and non-contradictory, its meaning must be understood by the learner (intelligible), and it must be believable (plausible) and useful in solving other problems (fruitful). Thus, preconceptions and conceptions introduced through teaching are seen as competing in terms of status in regard to intelligibility, plausibility and fruitfulness, in a process mediated within the learner's epistemological commitments or conceptual ecology (Toulmin 1972); a term that was subsequently seen as encompassing a variety of aspects including metaphors, exemplars and images, past experiences and metaphysical beliefs (Strike and Posner 1992).

Within this constructivist notion built on Piagetian ideas, teaching practices are typically concerned with the identification of learners' conceptions, promotion of dissatisfaction with them in terms of their explanatory power, and the introduction of more intelligible, plausible and fruitful scientific explanation. There is some claim that these strategies support learning more effectively than traditional didactic approaches (Duit and Treagust 2003), although this is not without contention. The range of strategies employed in promoting conceptual change includes that of creating cognitive conflict, a process in which the explanatory limitations of a learner's existing conceptual framework are deliberately exposed through introducing experiences that challenge existing ideas. The creation of cognitive conflict in learning is a commonly used instructional strategy designed to promote conceptual change. In this process the promotion of disaffection with personal constructs is intended to bring into question the usefulness of learners' thinking. Anomalous or contradictory data (derived from observation, investigation or secondary sources) is introduced into teaching such that personal constructs are challenged with the assumption that this will pave the way for the introduction and adoption of the scientific concept. However, research indicates that in reality the process produces variable outcomes with some learners responding positively whilst others fail to recognise the conflict, ignore it, avoid responding to it or only partially resolve it using alternative conceptions (see for example, Gauld 1986; Dreyfus et al. 1990; Chi 1992; Chinn and Brewer 1993; Shepardson and Moje 1999; Niaz 1995; Trumper 1997; Tao and Gunstone 2000; Parker 2006). The potential use and limitations of cognitive conflict in promoting learning are explored further in Chapter 4.

2.2.3 Developing Knowledge and Understanding of Learners' Conceptions in Science

Integral to the process of conceptual change is the central notion of shift in perception from existing ideas based on intuitive, common sense reasoning towards scientific explanations. Of critical significance in this process is the role of learners' existing knowledge, understanding and experiences within the domain. During the 1970s and early 1980s studies began to emerge that accounted for students' science learning in terms of domain-specific factors, rather than Piagetian global logico-mathematical reasoning skills. Research describing learners' ideas proliferated, encompassing every domain of science to be encountered in the curriculum across all ages and cultures (see for example, Ausubel 1968; Nussbaum and Novak 1976; Driver and Easley 1978; Viennot 1979; Gilbert and Watts 1983; Osborne and Freyberg 1985; Driver 1989a; Vosniadou and Brewer 1992; Kruger et al. 1990; Trumper 2001a; Galili 2001). This research is ongoing and now there exists a bibliography of works that has been comprehensively collated since the late 1970s that currently amounts to over 7,000 published works (Duit 2008). Although the terminology relating to learners' existing ideas in science has been described variously as, for example, *preconceptions*, *ideas*, *conceptions*, *naïve conceptions*, *alternative frameworks* and *misconceptions*, commonalities can be identified:

- learners of all ages possess preconceptions or beliefs about the scientific phenomena and ideas they encounter in science education.
- learners' preconceptions are unlikely to concur with scientific explanation, indeed they may differ significantly from it.
- learners' preconceptions are frequently robust, deeply rooted and difficult to influence in teaching.
- common preconceptions have been found in learners across cultures and age groups

Learners' preconceptions in science do not constitute isolated facets of knowledge; rather they are combined into an overall structure with its own logic. This complex, structured logic serves as a frame of reference for understanding the world (including science taught in formal settings) and presents significant challenge in pedagogy. Vosnaidou (1994) regards such conceptions as integral parts of broader, more complex cognitive structures and others contend that conceptions have several 'facets' called *explanatory systems* that are activated according to context (Minstrell 1992; Palmer 2001). Rather than being viewed as barriers to be removed during learning, Vosnaidou considers learners' conceptions to have a degree of plasticity that makes them virtually impossible to displace completely but is of the view that such conceptions are modifiable in teaching. In reviewing conceptual change research, Duit and Treagust (2003) comment that:

> There appears to be no study which found that a particular student's conception could be completely extinguished and then replaced by the science view. Indeed most studies show that the same old ideas stay alive in particular contexts. Usually the best that could be achieved was a 'peripheral conceptual change' (Chinn and Brewer 1993) in that parts of the initial idea merge with parts of the new idea to form some sort of hybrid idea (Gilbert et al. 1982; Jung 1993) (Duit and Treagust 2003: 673)

2.2.4 Some Theoretical Models of Conceptual Change

Whilst the classical model of conceptual change derived from the work of Posner et al. focused on the *conditions* under which radical change might be promoted, developmental psychologists grappled with the difficult issue of the nature of conceptual organisation; *what* changes and *how* it changes in learning. This has given rise to a range of perspectives with different emphases ranging from a concern with the minutiae of conceptual categories towards broader notions of knowledge systems (Limón and Mason 2002). Exemplification of the former includes the differentiation between weak and strong conceptual change identified in the work of Duit and Treagust (2003) who infer a typology in which knowledge accretion is considered towards the lower end of the spectrum of change continuum. This parallels Bloom's (1956) taxonomy of cognition, a view supported in the work of Carey (1985) and Harrison and Treagust (2000). The significance of ontological categorisation (Keil 1979; Chi 1992; Chi et al. 1994) and Chi and Roscoe's (2002) view that conceptual change is a process of cognitive repair are further examples. The latter attributes the problem to miscategorised concepts, a condition that results in misconceptions. An example of this would be 'weight' being assigned to a category of 'property of an object' as opposed to the category of 'force'.

Alternative perspectives such as those proposed by DiSessa and Sherrin (1998) consider that the notion of 'concept' needs to be replaced by more carefully defined theoretical constructs within a knowledge system. DiSessa (2002) regards learners as organisers of complex knowledge systems (*conceptual ecology*) in which knowledge becomes organised from the fragmented to the structured; the mechanism for conceptual change is not a simple process of deletion or replacement but rather a complex cognitive process of integration and reorganisation. He identifies co-ordination classes that are cognitive strategies such as selecting and integrating information about the world and *phenomenological primitives* or *p-prims* that are abstractions from experience and form primitive schemata that constitute the basis of intuitive knowledge. An example of force as a mover is a p-prim that arises from the observation of pushing an object and watching it slide across a table in the direction of the push (Clement 1982).

Variation within this second category is offered by Vosnaidou (1994) who considers concepts as being embedded into larger theoretical structures of two types. *Theoretical frameworks* develop from early infancy and are composed of fundamental ontological and epistemological suppositions whereas *specific frameworks* either involve belief about the properties of the behaviour of objects that arise from observation or are transmitted culturally. These theories provide the basis for the generation of specific mental models in response to the demands of a particular situation. Vosnaidou depicts conceptual change as a process of synthesis whereby learners seek to build a coherent explanatory framework as they attempt to reconcile inconsistent explanatory models by integrating material from new science experiences with existing frameworks (Vosnaidou 2002). This is a continuous process that takes place gradually as already constructed knowledge is enriched or restructured. For instance, the opportunity to explore pushing an object across surfaces in involving minimal friction may help to reconcile the theoretical explanatory framework of force acting on objects with the intuitive explanation of force as an entity consumed in motion.

Other theorists subscribe to the notion of intuitive mental rules that apply in principle across contexts and are stable and resistant to change (Stavy and Tirosh 2000). Matthews (2000) considers that there are innate dispositions to interpreting the world in particular ways with neural networks designed to respond quickly reinforcing an initial bias. In this paradigm, conceptual change is perceived as the building of additional cognitive structures that can override the functioning of competing innate structures. Although a universal theory of conceptual organisation and change remains elusive, what emerge from such research are a picture of complex mental organisation and a process of change that serves to emphasise the robustness of learners' preconceptions and the challenge of achieving change in teaching and learning science.

2.2.5 Considering the Individual's World

Latterly some limitations of the classical model of conceptual change have been highlighted (for example, see Pintrich et al. 1993; Vosnaidou and Ioannides 1998; Fensham 2001; Duit and Treagust 2003; Sinatra and Pintrich 2003; Scott et al. 2007). Important influences that appear to be closely connected to the learning of science concepts have been identified and, since the mid-1980s, research into the meta-cognitive level has encompassed learners' views on the nature of science and epistemological beliefs. Conceptual change concerns more than just the concepts of the learner and those of science. Sinatra and Pintrich (2003) emphasise the role of learners' intentions within the learning process and claim that as a consequence, the impetus for change resides with the learner and the notion of conceptual change becomes intentional. Furthermore, traditional approaches, described as being essentially rational or 'cold', focus only on changes to the understanding of subject matter, and fail to take adequate account of the influence of both the learner and the learning environment (Pintrich et al. 1993; Fensham 2001). Vosniadou and Ioannides (1998) draw attention to the development of metaconceptual awareness by students within the conceptual change process:

> Increased awareness of one's own beliefs and presuppositions and of the fact that they represent interpretations of physical reality that are hypothetical and can be subject to empirical test, is a necessary step in the process of conceptual change. (Vosniadou and Ioannides 1998: 1227)

A key component of the understanding of conceptual change processes in recent years stems from the recognition that learning cannot be extricated from the socio-cultural influences of the learning environment and that although theoretical frameworks have been described variously in the literature, they do not constitute a complete explanation of how students learn science in classrooms (Leach and Scott 2003). This focus on the microcosm of the learner's world emphasises the importance of affective influences in learning (Sinatra and Pintrich 2003). Learner attitude, for instance, has an important influence in the learning process, and a range of attitudes may impact upon an individual's motivation to learn science, such as the context of learning or how a person perceives their learning environment (Hodson and Hodson 1998). Motivation to learn is in turn influenced by students' self-efficacy, goals, intentions, beliefs, expectations and needs (Pintrich et al. 1993), and these are mediated within the physical, social and emotional parameters of the classroom. Chin and Brown (2000) discuss how research has identified a range of factors both personal (for example, cognitive style, motivation) and situational (for example, context, teaching methods) factors that influence a learner's approach.

Central to cognitive approaches of the 1980s and 1990s is the notion that an individual's knowledge and understanding of the world is constructed rather than received or transferred and this places importance on eliciting and responding to learners' thinking in various domains in the quest to design and evaluate teaching sequences to promote conceptual development (see for example, Stavy and Berkowitz

1980; Clement 1993; Minstrell 1992; Driver et al. 1994; Adey and Shayer 1994; Viennot and Rainsom 1999). Whilst the interpretation of the term constructivism is multifaceted (Strike 1987; Phillips 2000), and the extent and usefulness to which constructivism can be considered a theory of learning, teaching or both has been questioned (Solomon 1994; Fox 2001), it has, nevertheless, served to focus awareness of children's ideas and promoted teaching that seeks ways to challenge thinking through scientific inquiry, as learners actively construct their own understanding of the world as a result of their experiences and interactions (for example, Harlen 2000; Sharp et al. 2000).

The socio-cultural tradition, rooted in the work of Vygotsky (1978), focuses on the learning context as an integral part of the learning process with learning conceptualised in terms of passage from the social context to the construction of individual understanding. Interactions take place on the social plane between teachers, adults and students, and provide the context in which individuals rehearse ideas through a variety of modes of communication including talk, gestures, writing, visual imagery, and enquiry (the tools used by individuals for thinking). The social plane provides the forum for the meeting of new ideas and it is through this medium that individuals reflect on, and make sense of, ideas in the transition from the social to the individual plane. The socio-cultural perspective provides a major challenge to the tradition of mental structures (Vygotsky 1978; Wertsch 1991) as it represents a view of learning derived from social interaction. In the social sphere, language, the most important semiotic resource, provides the means by which individuals explore ideas and its fundamental role in learning science is discussed further in Chapter 5. Internalisation is a process whereby individuals develop and become able to use physical, conceptual and discursive tools within the social context. It is a process of individual sense-making rather than one of simple transfer. Conceptual change is perceived not as an individual process, but as the result of interaction between learner, tools and other people. Thus, the nature of the socio-cultural environment and the extent to which it encourages discussion of, and engagement with, personal and scientific ideas are important considerations in the learning of science.

Contemporary approaches to promoting learning drawing on Vygotsky's socio-cultural theory are often referred to as social constructivist approaches (see for example, Driver et al. 1994; Hodson and Hodson 1998; Wells 1999; Leach and Scott 2003; Mortimer and Scott 2003). Central to the social constructivist view is the notion that areas of knowledge, such as science, are a social product, developed within social communities and arising from social practices and that they cannot be learned through interactions within the world alone (see Driver et al. 1994; Leach and Scott 2003, for discussion of differences between empiricist interpretations of personal construction and socio-constructivist accounts of learning). This is not to say that individual interactions within the world do not play an important part in learning, but that learners also need to be given access to the concepts and models of conventional science, a process that takes place on the social plane and employs practices, conventions and modes of expression that are socially and institutionally agreed upon (Scott et al. 2007).

Thus, the nature of the socio-cultural environment and the extent to which it encourages engagement with personal and scientific ideas play an important role in the learning of science. In line with an international movement towards pupil empowerment and self-management of learning instructional practices have arisen that place emphasis on the enhancement of thinking skills (McGuinness 1999). This is manifest in the Cognitive Acceleration through Science Education project (CASE); an approach based on developing pupils' general intellectual functions that underlie context-independent components of general thinking (Adey et al. 1995; Shayer and Adey 2002). The initiative claims to achieve cognitive enhancement through moving pupils through the Piagetian stages of cognitive development at an accelerated rate in order to close the gap between curriculum content and cognitive demand. The CASE approach has produced positive effects in achievements, even in children as young as 5–7 years (Cattle and Howie 2008).

The implications of such a view of learning are that the teacher plays a crucial role in guiding the social discourse of the classroom to support and introduce scientific knowledge and explanation (Ogborn et al. 1996; Mortimer and Scott 2003; Chin 2006; Scott et al. 2007). This presents considerable challenge for teacher education in supporting both teachers and pre-service teachers as there is research evidence to suggest that teachers' views of science, and the teaching and learning of science, are often limited and restricted to the notion of science teaching as the transmission of factual knowledge (Parker and Spink 1997; Lunn 2002a, b). Trumper and Gorsky (1997) comment that many pre-service teachers hold the strong belief that good teaching is explaining through lecturing and this arises from student perceptions of their own educational experiences (Wubbels 1992). This view contrasts with notions of education being primarily concerned with developing understanding of science. In a study of pre-service teachers in initial teacher education (ITE), past experiences of learning in science were shown to be highly influential in developing attitudes to teaching science, especially when individuals perceive themselves as 'failed learners' who found it difficult to understand the explanations science has to offer:

> For me, the definition of science was purely physics and chemistry, both of which I found incredibly difficult and had many negative experiences with. My general ideas of physics and chemistry were scientific symbols and impossible equations, which I failed to solve, however hard I tried. (Parker and Spink 1997: 2)

The following examples of primary teachers learning about forces illustrate various aspects of the previous discussion and address the important challenge of promoting confidence in, and enthusiasm for, learning and teaching science. The account documents the teachers' endeavours to extend and develop both their science subject knowledge and pedagogic insight into the teaching and learning of the subject. It emphasises the core issue of meaning-making in science and exemplifies the conceptual change process in action as it occurs dynamically within the crucible of the classroom.

2.3 Conceptual Change in Action: Primary Teachers Learning About Forces

2.3.1 Forces Within the Context of Floating and Sinking

Learning about floating and sinking is a traditional science activity in English primary schools. In the early years of education, it often involves presenting children with tactile experiences explored through play. In later primary education, it provides an appropriate context for children to develop their ideas further. Learning includes recognising forces as pushes and pulls, identifying the forces acting on objects and considering their relative size and direction. Such ideas play an important role in underpinning understanding of balanced and unbalanced forces, a concept currently encountered in their early secondary education in England, but frequently implicit within in primary schemes of work. The challenging nature of this subject matter is such that some teachers may find themselves in the position of having to plan and teach the same concepts that they themselves struggle to understand and apply (Kruger et al. 1990).

The following discussion draws on a study involving pre-service teachers undertaking a one-year course leading to a Postgraduate Certificate in Education (PGCE) and a group of 30 primary teachers participating in a course designed to enhance teacher subject knowledge and understanding in order to support the effective meeting of the national curriculum requirements for science in the primary classroom (see Parker and Heywood 2000 for details). The principle objective of the research was to gain insight into how the teachers learn within a cognitive framework of balanced and unbalanced forces, using the context of floating and sinking. The background knowledge and experience of participants varied, with PGCE students with a range of first-degree expertise and primary teachers with experience of various levels of teaching within the primary sector. Although most of the teachers had little experience of forces in their former education, they had, nonetheless, all experienced some teaching relating to floating and sinking during their school careers.

The study aimed to identify how learners' thinking was influenced as they undertook a variety of activities based on floating and sinking. The activities were designed in order to:

- provide opportunities for experiencing the abstract notion of forces in a *physical* way
- elicit and develop learners' intuitive understanding of floating and sinking
- help learners to generate ideas and explore their own thinking within a practical context

The methodological approach to researching the teaching and learning, described in the introductory chapter of this book, is predicated on inquiry-based practice combined with strategies to promote metacognitive awareness of learning such that teachers and pre-service teachers (referred to as students or learners in the following illustration) articulate their thinking during the university-based teaching experiences. In charting their thinking as they make sense of learning experience, students become aware of the nature of learning and the challenges posed by the subject matter in question. This approach is contingent on the development of professional 'trust' between learner and teacher where the role of the tutor is one of facilitating the engagement of students with the subject matter through the activities devised to promote inquiry, questioning and discussion.

The research process deployed simple classroom activities to structure learning about forces and the sequence of instruction and learning intentions are summarised in Fig. 2.1. Activities ranged from open-ended exploration (activity 2) to structured systematic investigation (activity 4). An integral element of the process involved using strategies to challenge learners' ideas and expectations (activity 3). The notion of *weight in relation to size* was used as a scaffolding concept to help learners rationalise existing knowledge, understanding and experience in regard to floating and sinking. It also acted as a bridging concept leading towards the idea of density; an idea students sometimes associated with the context. Our experience of working with teachers in this area is that weight is a central tenet of adult learners' reasoning, and although it is often readily acknowledged that weight cannot be the sole determining factor in floating and sinking, so deeply embedded is this intuition that learners often find it difficult to comprehend the notion relationally to other determining factors.

2.3.2 The Socio-Cultural Environment and the Role of the Tutor

Creating an environment in which learners feel comfortable and confident in sharing their ideas with colleagues and tutors must not be underestimated as a pivotal aspect of such an approach to learning. This is a complex matter that, in addition to the creation of a comfortable physical environment with appropriate access to resources, also concerns the affective dimensions of learning such as motivation and willingness to engage with subject matter. For instance, group dynamics, individuals' perceptions of the tutor and the tutor's role, appropriateness of learning activities and discourse concerning the nature of learning in science are all significant features. The metacognitive approach adopted here attempts to make explicit the development of science knowledge and understanding through the deliberate 'problematising' of subject to support the realisation that all learners are on a continuum of understanding. It depends on the skill of the tutor in acting as a facilitator, helping to orientate student thinking about (in this case) the forces acting in a variety of contexts. This is illustrated in the following examples of learning presented in this chapter as the tutor engages learners in discussion about concepts such as: forces as pushes and pulls, gravity, weight, mass, upthrust and balanced/unbalanced forces.

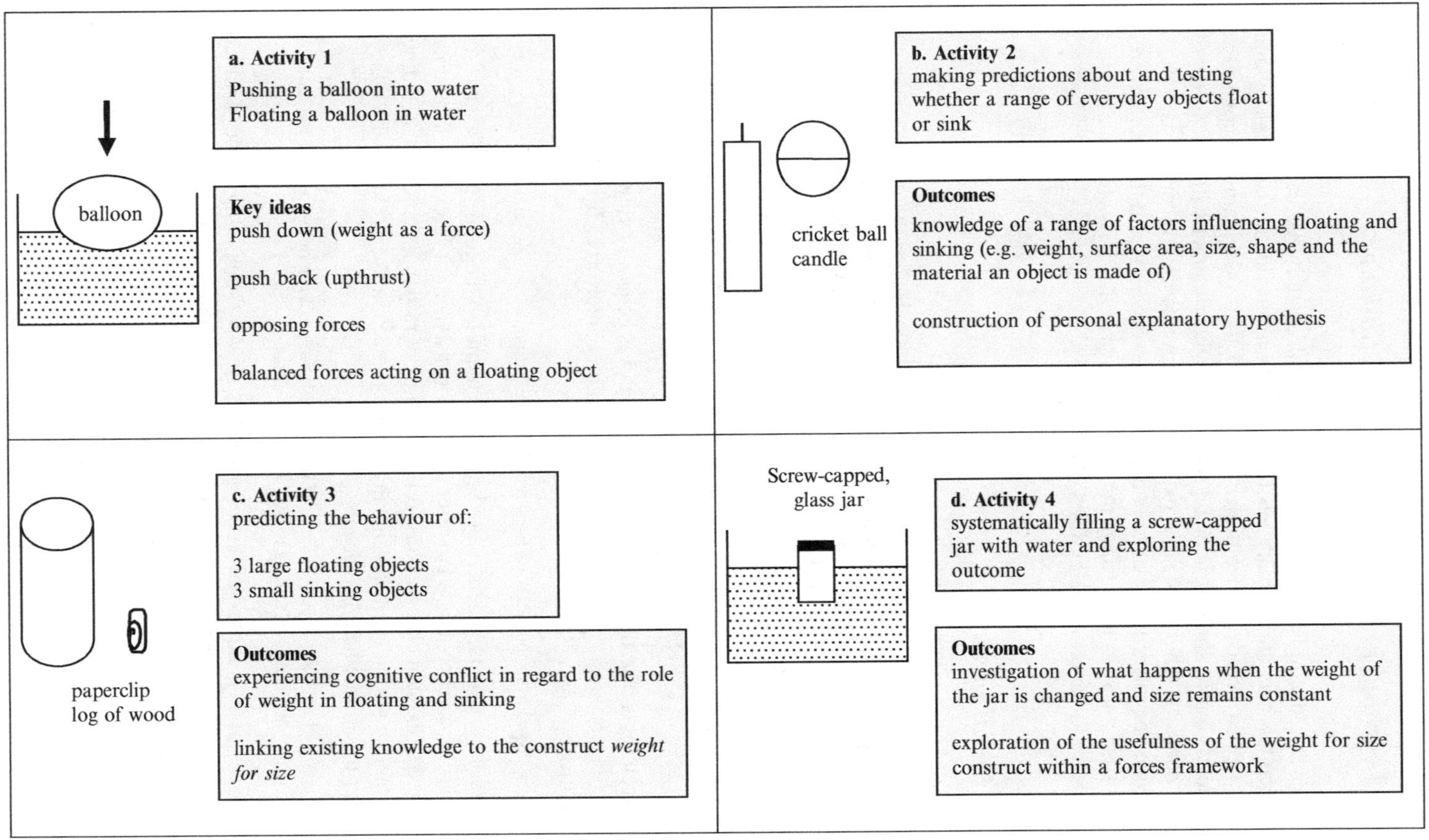

Fig. 2.1 Sequence of instruction in exploring floating and sinking

2.3.3 Learning in Action: Floating and Sinking

The following examples are drawn from a detailed study we carried out with pre-service teachers and primary teachers learning about floating and sinking and quotations presented derive from their accounts of their own learning captured at critical points during the instructional sequence as they interact within a variety of learning contexts. There are also extracts of conversations recorded as groups worked together and with the tutor in exploring the scenarios, and the data are presented in such a way as to illustrate the main findings of the study (for detailed data analysis refer to Parker and Heywood 2000).

2.3.4 Initial Ideas

The learners in this study demonstrated the common frameworks of thinking in regard to forces described in the introduction to this chapter and these are evidenced in their explanations for floating and sinking. By focusing on immediate responses to the first activity of pushing the balloon into water (Fig. 2.1a), we can gain a sense of how they intuitively rationalise the experience and express uncertainties in explaining what they think is happening. The physical experience of pushing a balloon into a tank of water generated considerable surprise even amongst experienced teachers:

> It was surprising to feel such a strong resistance from such a light object.

The seemingly puzzling observation, that what was generally viewed as a 'weak' material (water), was able to exert such a strong counterforce, frequently caused consternation:

> Is the resistance from the water or from the air in the balloon?

In describing events, teachers used the curriculum language of 'force', whereas the pre-service teachers tended to refer to 'water resistance' or 'pressure'. On being encouraged to explain observations further, a range of responses emerged including the Aristotelian notion of objects having a natural place in the world to which they endeavour to return when displaced:

> It [*the balloon*] naturally wants to float, in other words we are attempting to force the balloon to sink when if it is left alone it would naturally float.

Learners frequently expressed knowledge that it was 'something to do with' a variety of factors, but found it difficult to be explicit about the nature of the relationship between them, exhibiting an almost intuitive feel for the situation, possibly applying intuitive mental rules (Stavy and Tirosh 2000). Initial thinking takes time to formulate and practical activity offers a medium conducive to making observations that confirm, reinforce or extend previous knowledge. It stimulates the processing of information that relates to the mental organization of concepts through which the information is interpreted and which can ultimately result in articulation of thinking and the ability to express ideas. Discussion and the need to explain reasoning provide the opportunity not only to refine thinking through sharing and comparing ideas, but also through social interaction to help learners become aware of where uncertainties lie. Our research has shown that this is not a straightforward process and explanations are not always apparent to learners themselves, nor are they easily expressed. Initial explanations are often tentative and frequently begin with 'it's something to do with' as learners try to recall previous knowledge and experience but struggle to make useful connections with the present context:

> I do know this but I can't quite recall it.
>
> It's something to do with density and surface area
> It must be something to do with the amount of water displaced.

Experienced teachers were more comfortable in reasoning within a forces paradigm:

> The volume of the balloon displaces the water and the force of the water pushes the balloon back to the surface.

In writing and discussion, teachers commonly subscribed to the notion of upthrust as an entity residing in or belonging to the water. This is implicit in the statement above where the student refers to 'the force of' the water. As we shall see in this chapter, this idea comprises a considerable conceptual hurdle in learning; it is a robust construct that has a powerful influence in thinking about force and motion. For the purposes of this discussion we refer to the construct as the *entity view of force*.

Activity 2 (Fig. 2.1b) allowed learners to apply their wealth of life experience in making predictions about the floating or sinking of a range of everyday objects. Where teaching provides opportunity to apply such experience, individuals usually respond positively and confidently and, in this case, they were able to identify a host of factors influencing floating and sinking. Weight, shape, surface area, air content, density and the material from which the object was made, all featured prominently in thinking. Intuitive thinking tended to focus more on the shape or surface area of an object rather than its size or volume (the parameters necessary in constructing an understanding of density). A coherent understanding of density requires that two properties of an object (mass and volume) be held together in a relationship. By looking closely at the formulation of hypotheses, we found that few adult learners combined two or more factors in reasoning about floating and sinking. Even where they appeared to be employing density (based on mass per unit volume), this was applied to the material from which the object was made, as opposed to the object itself. It is important to note that learners do not instinctively engage in reasoning within a forces framework in contemplating floating and sinking and yet this is a context widely employed in primary schools to foster young children's understanding. If learners are to be successful in this enterprise they must be supported in developing a forces rationale for the context and teachers will need to consider the implications for children's learning.

2.3.5 Constructing and Reviewing Hypotheses

Activity 3 (Fig. 2.1c) was designed specifically to challenge and stimulate discussion of the central notion of weight by exposing learners to cognitive conflict in the form of large, heavy floating objects and small, light sinking objects. The intuitive influence of weight is rooted very deeply in most people's minds; in general heavy objects *do* sink. Even following the observation of large, heavy floating objects and small, light sinking objects, approximately 45% of the group continued to employ weight as the single determinant. It may be that this is an example of Stavy and Tirosh's (2000) intuitive rules about how the world operates and clearly, recognising that heavy objects often sink may have survival value in terms of how organisms live successfully in the otherwise precarious natural environment. Where surface area is used in reasoning, it tends to be underpinned by the entity view of force, in that the greater the surface area of an object, the larger the area over which upthrust (residing in the water) can act and the greater the resultant upward force:

> The bigger the surface area, the more the water's force can act on it.

This is a well-known and common robust conception and one that is more often than not confirmed by everyday experience of the world (Watts 1983; Brown and Clement 1987).

Following destabilization of thinking through the introduction of the cognitive conflict presented by activity 3 (Fig. 2.1), it is essential that learners are supported in resolving the conflict through developing appropriate and sensible explanation. To this end the construct of weight for size was introduced in scaffolding thinking and to assist them in developing meaningful explanation for floating and sinking (Fig. 2.2).

The challenge presented by this construct should not be underestimated in learning; it requires that the learner holds in their mind not one, but two concepts (weight *and* size). Furthermore these concepts must be considered in relation to each other (weight *for* size) and the learner must be able to apply this relationship in explaining a wide variety of contexts if it is to form the basis of an intelligible, plausible and fruitful explanation. Figure 2.2 illustrates how this construct provides a framework for rationalising existing knowledge including density. Some used, for example, the construct in considering density in relation to how closely packed the molecules of materials were and began to incorporate air into their reasoning in regard to its particulate arrangement as a gas. Occasionally some found it difficult to move beyond the misconception that air makes objects float and adhered to an untestable hypothesis in maintaining that as the three small objects contained no air then

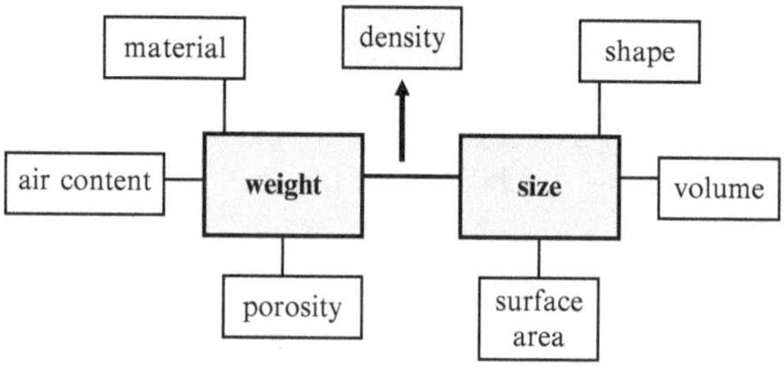

Fig. 2.2 A possible framework for linking the concepts of weight and size

the large floating objects must contain large amounts of air even if it could not be seen directly:

> Although you can't see it, there must be lots of air in the wax so that's why it can float.

The extent to which learners viewed air as a light material as opposed to an uplifting force was unclear, but this notion is also evident in younger learners (Driver et al. 1994) and written reflections did contain evidence of both:

> Air is buoyant, it makes things float up.
> Air is light because of its loose molecules.

2.3.6 Developing a Forces View of Floating and Sinking

The idea that weight is a force is critical in building a forces explanation of floating and sinking. Although the students in this study appeared to accept the idea readily, when individual reasoning was scrutinised, we found that most employed weight in the capacity of 'heaviness' (a property of the object). Even where they engaged in using the idea that weight is a force in conversation, their reasoning and reflective writing was frequently at odds with this perspective. As a consequence, learners must be supported in making the transition to a forces view of floating and sinking and activity 4 (Fig. 2.1d) was designed to facilitate this process. It provided an opportunity to systematically change the weight of an object (a small glass jar with a screw-capped lid), whilst keeping the volume constant. There were three emergent characteristic responses to this activity:

1. Making relevant observations and noting salient features that stimulate thinking:

   > It reaches a critical point where adding more water makes it sink.
   > The jar falls on its side and doesn't float upright.

2. Asking generalised why questions:

   > Why does it just get to one point then sink?

   Such responses are not untypical of initial thinking in new contexts (Parker 2006) and they tend to arise when an aspect of a phenomenon that catches the attention is not immediately explainable to a satisfactory degree on a personal level:

   > Why is the jar floating like that?
   > Why does it sink past a certain point?

3. Developing explanations:

   Learners' explanations are often tentative in the first instance and are frequently expressed in the form of questions:

   > Is it because there are two materials of different densities? (glass-jar, metal-lid)

   In focusing on the application of weight for size through adding water to the jar, we found that a critical point in learning emerged when students began to spontaneously engage in discussion of equal and opposite forces (weight and upthrust) acting in balance on a floating object:

The downward force is the weight and the water gives back an equal and opposite upthrust, these two forces are in balance when it floats in the water or rests on the bottom of the tank.

This represents a major conceptual shift in thinking as learners move from an intuitive perspective to one in which weight and upthrust are viewed as opposing forces with attendant effects on the motion of an object. The following conversation between the tutor (T) and two students (A and B) exemplifies this transition:

T: What's happening to the jar? *[students adding water to the jar]*
A: Gets to the point where it sinks.
T: So why do you think that happens?
B: It's (the water) making it heavier . . . cos there's less air in it and more water, so it's heavy.
T: Has anything else changed?
A: No, it's the same, the jar and everything; it's just got more weight with the water.
T: So is it just the weight that makes things float or sink?
B: No it can't be cos the big jar floated didn't it?
A: But if you put water in it, it would sink eventually
T: So what can you say about it then?
B: It just gets too heavy . . . for its size.
A: Yes, it must be both things really.
T: What can you say about the forces?
B: Well it [*the weight*] just gets bigger and the push back [*upthrust*] can't keep it up anymore.

Introduction of the construct of weight for size precipitated the conceptual shift and enabled learners to rationalise intuitive thinking. A student, for example, working on the initial premise that weight is the determining factor and other influences such as size, shape, air content are important, was subsequently able to explain why these factors influence floating and sinking in terms of an object being either heavier or lighter for its size:

The trapped volume of air keeps the jar light for its size. When it's half full of water it's heavier-volume of air is reduced but still light for its size (density) so it floats lower in the water.

The activity also allowed a further opportunity to focus teaching on differentiating mass and weight (critical in formulating density), although these terms were far from clear in the minds of most learners at this stage:

When water is added its mass for size (density) is increased.

2.3.7 *Generalising Weight for Size*

If learning is not to be restricted to the narrow context of the teaching activity, learners must have the opportunity to make links with wider experience. This involves trying out new concepts to see how useful they are in a wider sense.

We found this process occurred rapidly and spontaneously as students developed their understanding of balanced/unbalanced forces and the effects of changing the weight for size of the object in activity 4 (Fig. 2.1d). Thought experiments, such as what would happen if a vacuum was created inside the jar or the jar was floated in very salty water, stimulated this process:

> It's just like a submarine isn't it?
> It [*jar*] would float higher in salty water because the salty waters more dense!
> Salt water would support a heavier weight.
> Would heavy objects sink to the bottom of the Dead Sea?
> If we put helium in instead of air it would be lighter and float higher.

Thought experiments, although the product of mental activity, can be regarded as empirical experiments that could not be (or have not been) conducted (Sorensen 1992) and constitute an important goal in physics education (Gilbert and Reiner 2000). Reiner and Gilbert (2004) found in a study of secondary school pupils' learning about magnetism that the process of alternating between experimenting practically and experimenting in thought leads towards a convergence of scientifically acceptable concepts; they called this process mutual convergence. It seems that an integral part of making sense of the everyday physical world involves students generating and using their own spontaneous thought experiments (Reiner 1997). Being able to generalise an explanation successfully in this way is a very positive, motivating experience for both teacher and learner alike. It represents the coming together of ideas for the first time for a particular learner and is what teachers continually strive to achieve in practice.

2.3.8 Understanding Forces in Different Contexts – Towards Context Independent Learning

In seeking to develop effective and useful explanations for the behaviour of the living and non-living world, science prizes the development of theories and laws with wide explanatory power. In the learning of science, conceptual change research studies have demonstrated that conceptual change is likely to be restricted to the context of the task; learners frequently reverting to old, alternative concepts in different contexts (Tao and Gunstone 2000). Learners often vacillate between applying new and old concepts, and context independent change where commonalties are perceived across contexts seems to be difficult to achieve. Demastes et al. (1996) describe a spectrum of conceptual change patterns and note that students can even hold two logically incompatible conceptions at one time (dual construction). Similarly, Fensham et al. (1994) noted the existence of competing conceptions that coexist in learners' minds. Context-independent conceptual change may well be a slow and complex process, but it remains an ultimate goal of science education.

We conducted a subsequent study to explore how learners retained key ideas from the floating and sinking activities and to explore how these were applied to other contexts involving forces and motion (Heywood and Parker 2001). It focused on 18 PGCE students of varying educational background and experience, none of them were science specialists, and explored their learning in a variety of contexts involving stationary structures, and horizontal and vertical motion. Key ideas identified by the students as a result of having participated in the taught session of floating and sinking were:

- weight is a downward force.
- upthrust is exerted by the water.
- the forces acting on a floating object (weight and upthrust) are balanced.
- the forces acting on a sinking object (weight and upthrust) are unbalanced.
- an object's weight for size determines whether it will float or sink.
- a range of factors influence floating (for example, shape, material, air content).

Despite having considered gravitational attraction, weight and mass in thinking about floating and sinking, there was still a degree of confusion over these terms. In writing, for instance, and conversation, gravity and weight were often used interchangeably and, at other times were clearly represented as two separate and distinct forces:

> Gravity (weight) is acting downwards.
> The gravity of the Earth gives the object weight.

The intuitive, entity view of forces belonging to objects was still evident in some of the group:

> The water **has** an upthrust.
> The bigger the surface area the more **the water's upthrust** can act on it.
> [*author emphasis*]

Although the concept that weight is a force was readily identified as a key idea in learning, when it came to applying it to the context of a simple arched bridge, we found that learners frequently failed to transfer the notion. Also, although the English national curriculum requires that primary children be taught about 'gravitational attraction' between objects, in discussion and writing this term was rarely used by adults who tended to describe 'gravity' as a downward or pulling force emanating from the Earth and referred to this variously as the Earth's attraction/pull/force. While the scientific definition of weight concerns the gravitational attraction between masses, students' general reference to 'gravity' implies the existence of gravity as an independent force or 'entity'. We will return to this later as it has significant implications for understanding vertical motion. There was still a degree of uncertainty concerning the notion of weight, while appearing to accept the notion of weight as a force, in reality many reverted to the intuitive view of heaviness in reasoning about the context. Mass was generally considered as 'the amount of matter making up an object' or 'the amount of space/surface area of an object'; unlike weight, it was used infrequently in spontaneous social discourse.

In order to explore how learners transferred concepts to different contexts, we will look at their subsequent thinking in two contrasting contexts. The first context requires consideration of the forces acting on stationary structures and the second focuses on vertical motion.

2.3.9 *The Arched Bridge*

A simple arched bridge (constructed by bending a stiff, plastic sheet between two small block supports) was systematically loaded and students were asked to reason about the forces acting as the bridge was gradually loaded leading ultimately to the point of collapse. Interestingly, they found it easiest to reason about the bridge when it was carrying a load, some failing to recognise the weight of the unloaded bridge as a downward force. They were quick to draw parallels between the arched bridge and floating/sinking and these are summarised below:

Floating and sinking	**Simple arched bridge**
Weight of object acting downward on water	Load acting downward on bridge
Floating object remains at surface (forces balanced)	Loaded bridge does not collapse (forces balanced)
Increasing weight of object causes sinking (forces become unbalanced)	Increasing load causes bridge to collapse (forces become unbalanced)

Their explanatory diagrams illustrated how they readily applied notions of opposing forces acting within the structure (see Fig. 2.3 for an example) and they postulated that as the bridge did not collapse, the forces must ultimately be balanced by the weight of the supporting blocks acting on the ground. This represents a considerable conceptual shift for those initially subscribing to the notion that there are no forces acting on stationary structures.

Although learners are quick to pick up on the similarities between the floating and sinking and arched bridge scenarios, it is the differences between the contexts that invariably present barriers to conceptual transfer. The students, for example, found it easy to reason about the bridge when it was carrying a load (often referred to as a 'weight') as they could equate this with a floating object. However, when the bridge was unloaded they struggled to identify a downward acting force. Similarly, the source of upthrust is more tangible in floating and sinking as learners can feel the force being exerted by the water. However, in seeking to apply a framework of opposing forces to the bridge, several students sought to pinpoint the source of 'upthrust' in the scenario:

> What is the 'push back' under the arch?
> Is this push back the upthrust of air?
> Is it only water that exerts an upthrust?

In identifying an opposing force a conceptual dilemma arises in that a 'downward force' (load) is balanced by an apparently 'sideways' acting force (Fig. 2.3). The systematic loading of the bridge also generated a conceptual problem for learners developing a coherent, causal mechanism to explain why the forces remain in balance, despite the increased loading:

> Why are the forces still in balance? How does it know to counterbalance the increase?
> I'm not sure if the push back from the brick has increased . . . it's too difficult to imagine!

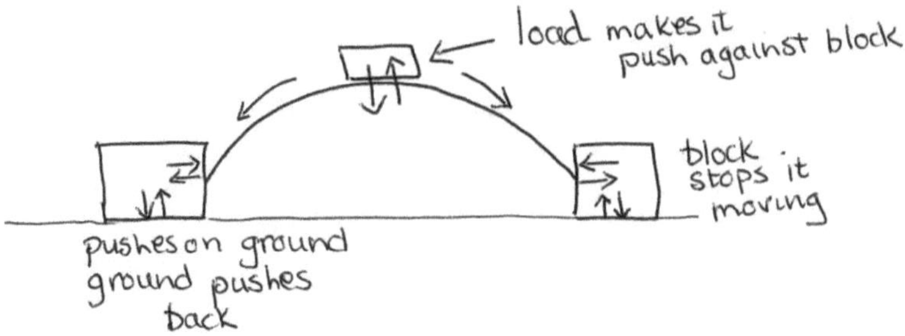

Fig. 2.3 An example of a student's explanatory diagram of forces acting on a simple arched bridge

In essence this is the same problem faced in rationalising the floating of objects as their mass is increased. Its resolution depends on the movement from an entity view of force towards the Newtonian view of action–reaction. The difficulty of this is embodied in student response to the activity in that some simply confirmed and reinforced their intuitive notion that a quantitative amount of force must reside in the supports:

> There is still enough force pushing back to hold the object in place.
> The downward force of the brick (load) is not heavy enough to affect the force of the brick holding it in place.

The challenging nature of this scenario generated a plethora of uncertainties indicating a degree of destabilisation in thinking:

> I'm unsure how forces get transmitted sideways.
> I'm confused about what the downward forces are.

In a sense the bridge acts as a black box in that while force is transmitted within the structure of the material from which the bridge is made, the process is not visible to the observer. Once again reasoning illustrated the robustness of the entity view of force as they raised questions concerning the generation of opposing force:

> How does the force in the bricks compensate? Where does the extra push come from?

This kind of thinking can have one of two outcomes; learners may find the problem too daunting and simply cease to pursue it further, or they can be driven to resolve the issue through pursuing a different line of investigation. In this study some individuals began to think about the nature of the materials used to make the bridge. With tutor help they contemplated, in a very simple way, the particle theory of matter and this provided a tool that could be used in formulating rational explanation. In considering that particles are bonded, albeit variously, they reasoned that a force applied to a bond would result in a 'push back'. This could be related to concrete examples of how springs behave when forces are applied to them and students found this a useful way of thinking about the transmission of forces; a similar analogy is used by Gordon (1978) in explaining the properties of strong materials. In focusing on the differences between the contexts we present an opportunity to extend learning and influence conceptual frameworks.

Akerson et al. (2000) suggest that analogies can be used to reduce the activation of erroneous explanatory systems and extend the domain of scientific explanation as they point out similarities between a situation that the learner can understand and knows how to explain (in this study the behaviour of springs/bonding between particles) and a new situation (transmission of force within a structure) with resultant transfer of knowledge to the new situation.

In reasoning about the arched bridge, the downward force acting on the bridge was referred to variously as weight, gravity and, weight and gravity, again revealing lack of clarity in thinking:

> Is gravity the downward force or is it weight?
> Is weight or gravity pushing it down, is there a difference?

Comments such as these are important as they indicate that learners are beginning to recognise that the words we use in general conversation may be at odds with thinking. The fact that this recognition emerges from observations made by the learners themselves creates an opportunity to discuss and develop meaning for these terms.

2.3.10 *The Parachutist*

The example of a parachutist's descent from an aircraft is a classic context employed in teaching and learning about forces. It provides a vehicle in which to embed ideas such as: air resistance and weight as forces, the nature of movement (acceleration/deceleration/qualitative distinctions between speed, velocity, constant speed and terminal velocity), as well as opposing and balanced/unbalanced forces. The following example of learning is focused on a small group of five pre-service teachers all but one of which began by considering the context in terms of gravity (as opposed to weight) and air resistance. The remaining group member discussed the context in terms of weight. Changes in motion were easy to detect in this scenario, especially when the parachute opened during the descent producing a dramatic deceleration. Although, at one level, the students were able to reason confidently about balancing and unbalancing effects on motion of weight/gravity and air resistance as the parachutist accelerates and decelerates during the descent, the notion of balanced forces acting on objects moving at constant velocity remained both puzzling and counterintuitive:

> The forces are in balance but it's still moving?
> Are you slowing down to a full stop eventually or are you at constant speed forever?

One member of the group had vague recollections of reading about Galileo dropping cannonballs from the Tower of Pisa and this pre-empted an interesting group discussion of what happens when objects of different masses are dropped from the same height. This is a particularly challenging concept and in order to support the students in thinking about it, two small containers of the same size (same air resistance) but different masses (one 10 times the mass of the other) were used.

Prior to dropping the masses all learners confidently predicted that the heaviest would hit the ground first. On reflecting on outcomes, much of the discussion centred on how 'gravity' acted on the two masses:

> Is gravity acting greater on the heaviest mass?
> I know it's something to do with gravity gives weight, but gravity acts the same on all objects.
> I think that gravity is acting the same on both objects (as it acts on any object on the Earth). The heaviest mass will fall quicker as the force that pulls it down (weight) is greater than the force exerted by air resistance.

Underpinning these observations is the notion that gravity is a constant force emanating from the Earth and acting the same on all objects. The issue of weight as force is once again manifest as a complication. Considerable conceptual disturbance (cognitive conflict) was created on discovering that the masses reach the ground at the same time:

> I still feel that the heaviest will hit the ground first, but I know it's wrong.

An example is provided of the tutor (T) helping a student (S) to resolve the conflict the by considering a situation she felt comfortable with:

> T: What is it that's concerning you?
> S: Well, I just feel that heavier things fall fastest.
> T: It might help to think about things moving horizontally first. If you were to move two objects of the same size but different masses on the same surface what could you say about the force needed to move them?
> S: The heavier one would have to be pushed harder . . . more force.
> T: Now think of the objects not being moved horizontally but moving vertically through the air, falling to Earth. What could you say about the forces?
> S: Well they'll just fall won't they? You don't have to think about the force they just do because of gravity.
> T: Mm, but what if there was no gravity and you had to move the objects?
> S: Well . . . you'd have to pull the heavier one more wouldn't you?
> T: Why?
> S: Because it's heavier, it's got more matter, there's more to move.
> T: So if it was 10 times heavier you'd need 10 times the force to move it?
> S: Yes.
> T: So what is the downward force acting to move the object?
> S: Gravity . . . but it acts the same doesn't it?
> T: Think about holding two different masses in your hand, how do they feel?
> S: No I see . . . they have different weight cos weight's caused by gravity acting on them. I see it's got 10 times the weight but it needs that to move it so they move the same!

Later the same student wrote in reflection:

> The key for me is what does it need to move it. You don't think when something falls it needs to be moved, it just does. But heavier objects will need more force to move them . . . I see it now.

In considering vertical motion, we found that learners tend to focus on 'gravity' as opposed to weight (as a force). The context serves to channel thinking in this way in that visualising the descent involves thinking about the Earth as a massive body in relation to the small falling object (parachutist). It was this difference in context that stimulated discussion in which learners began to differentiate gravity, weight and mass:

> I saw the weight and gravity as two different things (gravity as a constant + earth's attraction, full stop and weight as the force that makes an object fall quicker and slower).

> Weight is a consequence of the pull of gravity on our mass. Pulls object with a greater mass with a greater force. Hits the ground at the same time because it pulls harder on the one with more mass.

The depth of thinking needed for adults to begin to make sense of the core concept that weight is a force should not be underestimated in teaching and learning. This construct presents substantial conceptual challenge and one of the reasons forces remains a difficult subject in later education is that qualitative, meaningful explanation is not developed prior to quantitative application. The account of both pre-service teachers and teachers contending with the idea in different contexts is testament to this, and it is interesting to note that although students were beginning to differentiate weight as the result of gravitational attraction, there are still elements of their reflections that reveal that the notion of gravity as a 'property' of the Earth that acts on objects in its sphere remains an integral part of their mental representations.

2.4 Some Conclusions and Implications

2.4.1 Reflections on the Development a Qualitative Understanding of Force and Motion

In documenting the fine detail of learner interaction within the studies discussed, we could determine no distinct and predictable outcomes in terms of conceptual change on an individual basis. Despite careful tracking of individuals' progress in situ, it was not possible to predict the learning pathways and outcomes even for students beginning with the same initial subject-related conceptions and working in the same social groupings. They would not necessarily reach the same end point in learning or identify the same aspects of significance in their learning. This reflects the complexity of conceptual ecology and the socio-cultural environment within which learning takes place where individual sense-making is a personal process of unique and individual construction. This resonates with von Glasersfeld's (1992) view of constructivist learning in that teaching is a social activity but learning is a private activity with understanding being constructed by each individual 'knower'. The learning agenda is clearly set individually and the challenge for the teacher becomes one of how best to facilitate learning within a personal framework of reference. This is not to say that the participants in this study did not make progress in understanding key ideas, but the learning pathways taken by individuals varied as they contended with the challenges presented by the subject matter and they reached different levels of understanding. In scrutinising the minutiae of engagement with problematic scientific concepts, the teacher educator can develop valuable pedagogical insight into how the non-specialist can develop qualitative understanding in physics.

Finding out learners' conceptions, for instance, often presented as a relatively simple, routine affair in science education literature, can in reality prove to be a lengthy business that requires careful handling in the teaching/learning interface. This is to do with several factors, not least of which is the creation of circumstances in which adult learners feel secure enough to take the risk of exposing their reasoning within a social setting, or even in learning journals to be shared by student and tutor. Another consideration is the nature of intuitive thinking itself. Our studies show that intuitive thinking is often vague, combining elements of past experience and teaching that are recalled (often partially) as the learner attempts to make sense of a phenomenon. Elements of previous learning are frequently expressed as loose associations where ideas have been linked mentally, but not necessarily coherently, in providing a meaningful explanation that can be readily applied in reasoning. In floating and sinking, the use of density provides a classic case. Although all students in our study had knowledge of density, few were able to employ it in a useful way in personal reasoning. Intuitive constructs were most often based on common-sense reasoning, as illustrated in the example of Aristotelian reasoning, and the arguably common sense and ubiquitous, persistent entity view of force.

Learners' existing knowledge and understanding provides the lens through which learning is mediated. In providing the opportunity for learners to make predictions about whether a range of everyday objects will float or sink and to construct and explore personal hypotheses, we gained insight into how they made judgments and found that, in general, learners are unlikely to think intuitively about force or density. Despite having knowledge of the meaning of the terms from prior learning, the constructs do not inform the explanatory framework activated when it comes to reasoning intuitively about this context. If in teaching science we are attempting to engage learners in activating and employing ideas that are incommensurate with existing frameworks, then we must necessarily, as part of instruction, provide appropriate scaffolding to enable existing knowledge to be rationalised within the desired framework. This is particularly important pedagogical knowledge for primary teachers as they have the job of engaging young learners similarly in practice. The evidence drawn from the detail of the learners' perspective of the dynamic learning process can provide important insight. The tendency, for instance, to focus on surface area and shape in explaining floating and sinking, as opposed to size (volume), is likely to act as a barrier to the meaningful development of the concept of density at a later stage and consequently teaching needs to provide opportunity to engage with the relation between shape, size and surface area. The focus on surface area and shape derive from the underlying entity view of force. Indeed, the very fact that an object eventually sinks (i.e. is not supported by the water) serves to reinforce the notion that water contains a finite amount of force.

Learners struggle to comprehend that this 'entity' is not applied in full when an object exerts a downward force onto water.

To move from this position towards the idea that the force exerted by the water is a consequence of the interaction of the object with the water is not insignificant in further learning of forces. Brown (1989), in discussing a similar problem in the education of high school students, examines the consequences of learners failing to understand what is Newton's third law, epigrammatically described as 'for every action there is an equal and opposite reaction'. Critical in understanding the law is the recognition that a body can neither experience nor exert a force in isolation; that the force exerted by one body on another is exactly the same magnitude; that neither force precedes the other and that forces act in a direction exactly opposite to each other. Failure to grasp these concepts results in an inability to apply Newton's third law accurately in reasoning as exemplified by the learners in our study. Floating and sinking provides a classic example of this as the students' emergent questions show. Observations of the relationship between 'push down' and 'push back' invariably lead to the question: How does the water know how much force to apply? Newton's third law, however, only describes what happens; it does not satisfy the need for a *meaningful* explanation that is so important in securing conceptual change. We have found in practice that considering particle bonding, albeit in a simple way, can be helpful in supporting reasoning in that learners can readily conceptualise 'particles' of substances held together by 'bonds' represented by springs. They readily equate 'bonds' with 'springs' and can employ the analogy as a mechanism to explain how water or a bridge might transmit forces. However, like all analogies, if scrutinised too far it will ultimately cease to be helpful (see Chapter 3 for further discussion of analogical reasoning). It can, nonetheless, provide a useful bridging concept in explaining the action–reaction phenomenon.

The examples discussed in this chapter are testament to the cognitive challenge presented by the concept that weight is a force. Weight is most frequently viewed as a property of an object (heaviness) by adults and children alike and this presents an important potential stumbling block in physics education (Osborne and Freyberg 1985; Brown 1989; Trumper and Gorsky 1996; Bar et al. 1997; Galili 2001; Baldy and Aubert 2005). The only evidence in support of weight as a force that is in any way accessible to the learners at this level is the effect of reduced gravitational attraction between objects and the Moon or the phenomenon of weightlessness in space. This must be set against the multitude of everyday interactions in the world that serve to confirm the idea of weight as heaviness; we simply cannot demonstrate a concrete situation where weight changes that cannot be interpreted by the learner in terms of heaviness.

Baldy (2007), in considering the problem presented by the Newtonian view of weight as a force as a consequence of the attraction of masses at a distance, draws attention to the difficulties faced by French tenth grade students in conceptualising the force (is it magnetic or magic?). She points to the lack of coherent qualitative explanation that weight is a force combined with evidence drawn from everyday life where masses are not readily perceived as being attracted to each other, as important factors acting to reduce the intelligibility, plausibility and fruitfulness of the construct and thereby inhibiting conceptual change. Instead she proposes a geometric explanation based on Einstein's Theory of Relativity that does not rely on the force of attraction. In the so-called pillow model, a pillow represents space and steel masses of various sizes represent celestial bodies. If a marble is rolled fast enough it will deviate from its normal trajectory in the vicinity of a 'celestial body' thereby simulating 'falling'. This provides a concrete dimension to learning in this area that Baldy contends leads to improved understanding. Although a seemingly useful way of conceptualising falling masses on a personal level, this does not help the primary teacher who may well be faced with trying to develop young children's understanding of gravitational attraction in the early stages.

We found that the introduction and systematic investigation of the construct weight for size (Fig. 2.1 – activity 4) led to the spontaneous generation of thinking about forces and density. It provided the basis for thinking about the meaning of scientific terms such as mass, weight and density and their learning was evidenced in the proliferation of spontaneous thought experiments (including what might happen should the investigation be conducted on the Moon) and the making of links with existing knowledge (such as floating in the Dead Sea) as they sought to try out this qualitative explanation. It spawned a range of questions including asking whether there gravity on the Moon, indicating a large degree of uncertainty about gravitational attraction outside the sphere of the Earth that provided an opportunity to explore these issues further. Baldy and Aubert (2005) found that the 15-year olds in their study often relied on explanations of falling being dependent on the place where it occurs with bodies falling as a result of attraction being confined to the Earth as a special case whilst bodies were likely to float on the Moon or in space as there is no atmosphere.

Achieving context-independent conceptual change is a goal of education and without it, concepts remain bound to specific contexts and are, as we have illustrated, of limited use in reasoning. Some important insights emerge from following students' learning in different contexts. First, there is a need to proceed with some caution in making the assumption that concepts are fully embedded and that learners are able to generalise ideas and apply them to different contexts. Whilst they identified weight as a force as a key concept arising from their learning about floating and sinking, in considering other contexts they frequently reverted to an entity view of force and used gravity and weight interchangeably in conversation. Some students regarded falling as a natural process and failed to consider weight as a force, focusing instead on air friction as a force that slows descent. The failure to transfer the concept of weight as a force was evident in the context of the parachutist where weight is less tangible and the mental image of the falling body serves to focus the learner on 'gravity' as an attractive force emanating from the Earth.

Second, it is helpful in teaching to be clear about the differences between the contexts we present to learners. We can be tempted to ignore these in focusing on similarities, but in looking closely at personal reasoning, it is the differences between contexts that create barriers to the application of key ideas thereby reducing their usefulness as explanatory devices. In considering, for example, the arched bridge, individuals readily reasoned about balanced and unbalanced forces acting and confidently drew parallels with floating and sinking. However, this reasoning also served to precipitate uncertainty in that they thought that either the forces must be transmitted within the structure, or there must be a source of 'upthrust' to oppose the downward force. This is a critical point in learning in that where uncertainty is generated, opportunity emerges to extend learning. In this case students needed support in developing a coherent, causal explanation of how force can be transmitted within a material and structure. A consequence of identifying this question is that it created a stimulus to explore how bonding holds the particles of liquid and solid substances together and how this can account for the transmission of force. Students were content to conceptualise bonds as acting like springs or rubber bands in holding together particles of matter and found this a useful analogy in reasoning about the transmission of force. The important point here is that this learning stemmed directly from a consideration of the *differences* between the arched bridge and the floating and sinking contexts and students' awareness of the difficulties in transferring concepts.

In focusing on the differences between contexts, we can directly explore the problematic aspects of context that learners themselves have identified as conceptual barriers. In sharing the difficulties inherent in learning with learners we can overtly 'problematise' physics subject knowledge such that we expose and share with learners the unexpected, surprising and sometimes bizarre yet fascinating way in which physics describes how forces act.

2.4.2 Developing Pedagogical Insight Through Employing a Metacognitive Approach to Learning

In identifying and sharing observations on personal learning processes, students evolved valuable insight into the learning of not only subject but also the nature of inquiry-based learning in science and strategies that support it. They identified several general pedagogical aspects of learning, such as the role of discussion and influence of social interaction, but also focused on aspects of scientific investigation such as making predictions, formulating personal hypotheses and testing ideas that are of importance in scientific enquiry. Furthermore, they commented on the nature of their reaction to the learning environment such as how it feels when faced with cognitive conflict and the need to resolve conflict and construct a meaningful explanation for what is observed. Although described comprehensively in science education literature, these aspects of learning constitute original insight for the participants and provide valuable experiential contribution to emergent pedagogical knowledge.

Insights were also generated into the nature of the subject of force and the difficulties involved in developing an understanding of it. The abstract and counter-intuitive nature of forces, for example, was identified as important aspects that need to be addressed explicitly in teaching through providing opportunities for learners to have concrete, physical experience of force to relate their ideas to. They were able to recognise common misconceptions such as:

> Although air is important in floating and sinking, it does not actually make objects float.

> Floating and sinking cannot be explained by one factor only.

Furthermore, they articulated the precise nature of the difficulties encountered in coming to understand the subject better:

> I realised that when confronted with the question that big things are heavy and small things are light. I didn't know how to explore this further. In class we weighed things and identified that this was the case but I wasn't sure where to take it. The experiment we did today showed how weight for size is central . . . taking two things at once.

> For me, I never really thought about forces in something stationary and I think this is the case for children, but experiencing it has been very useful in my own understanding.

Developing metacognitive awareness of learning enabled teachers to appreciate something of the conceptual demand and pedagogical challenge of teaching and learning a subject where intuitive knowledge is extensive and highly influential in learners' thinking and the ideas to be taught are both abstract and counterintuitive. A particular concern was the need to build connections between learners' tacit knowledge and scientific explanation in meaningful ways:

> I now feel there' s a lot more to understanding floating and sinking than I first imagined. . . it's the building of ideas and putting them together . . . holding two ideas together is much harder than holding them both separately.

This is reflected in the work of Galili and Bar (1997) who explored the conceptions of weight held by children (aged 5–16 years), demonstrating that their views develop gradually from tactile experiences and are often constructed intuitively at a young age. They identified a range of schemes such as weight as a pressing force and weight belonging to heavy objects, the application of which could be mutually contradictory in different contexts. Significantly, they found that the identification of weight with gravitational force is not reached through natural development.

In adopting a metacognitive approach to learning, teachers are engaged in experiences that serve to generate empathy with the condition of being a learner of science and recognise the need to help learners to build effective qualitative explanations for their observations:

> I can understand how children might struggle with ideas.
> I will have to help them [*children*] to make clear hypotheses . . . otherwise lots of messy, pointless filling of jars will happen if there is no clarity of thought about what is happening.
> I have learnt the difference between qualitative and quantitative explanation.

This chapter has sharpened the distinction between knowing and developing understanding in science. The latter requires the construction and internalisation of coherent and causal explanations phenomena observed. The pursuit of meaningful explanation lies at the heart of both the scientific endeavour and personal learning. Research literature describes various models of learners' explanations showing a progressive trend from precausal to causal explanations that provide the individual with a convincing rationale for the phenomenon under scrutiny (see for example, Solomon 1986; Metz 1991; Woodruff and Meyer 1997). Such causal explanations are highly valued and contain considerable explanatory power

(Gilbert et al. 1988) and are, therefore, more likely to lead towards context-independent conceptual change. As one teacher commented:

> I now have a much clearer understanding of things. I have learnt about forces in the past but have tried to understand them at a higher level without having developed the basic knowledge.

2.4.3 Some Implications for Teacher Education

As beliefs about teaching and learning drive decisions to do with teaching (Biggs 1987), developing an awareness of the nature of the subject and of learning is, necessarily, a crucial part of teacher education. Conceptual change, such as that described in this chapter, would imply a deep approach to learning that entails intrinsic motivation and interest in the context of the task with a focus on understanding. It contrasts sharply with a surface approach to learning based on the perception of the task as a demand with learning taking place by rôte (Marton and Säljö 1976; Marton 1983; Biggs 1987). Learners must develop a convincing rationale for their observations of how the world works if they are to meet the criteria for satisfactory explanation and become part of personal reasoning strategies. Indeed, Dillon (1994), in discussing the value of qualitative reasoning in the learning of physics, points out that people seem to reason very successfully about physical systems where the only knowledge they have is qualitative. Qualitative reasoning concerns the skills used by expert practitioners to reason about the world in non-quantitative ways. In this chapter we have illustrated, from the learner's perspective, some of the inherent difficulties they experience in developing causal, qualitative explanations about forces in different contexts. It highlights Wolpert's contention expressed in the introduction to this chapter that science does not always accord with common sense and how, in order to understand and make sense of scientific explanation, learners must engage in significant depth with the subject. Science may not always fit with common sense but in order for the subject to have meaning, it must make sense to the learner and, furthermore, to the teacher who facilities learning in the classroom. While the teachers in the studies examined possessed knowledge of contexts and some understanding of force and motion, their explanations were unlikely to be coherent or transferable and they were not always able to activate their knowledge in productive ways in developing explanations. In teacher education, we have the opportunity to engage teachers in mapping the cognitive landscape of science domains such that they develop awareness of structuring learning and learners are supported in developing a coherence of explanation with which they feel comfortable. The teacher needs to be aware of those points in learning that are likely to be critical in introducing and/or reinforcing concepts. This would seem a prerequisite for context-independent conceptual change. However, learners need to recognise explicitly that integral to this process is the refining and reinterpretation of ideas such that they have meaning in different contexts. Existing ideas become a positive cognitive base domain for the projection of meaning in different contexts provided that they are coherent with respect to the specific phenomenon in question.

Chapter 3
The Role of Analogies in Learning

In this chapter, we first present an empirical account that documents teachers' learning about simple electric circuits through the use of analogies. In reviewing the analysis of data generated, we go on to propose that the research enterprise should shift focus from determining the effectiveness of analogy in cognitive transfer towards recognising the role of analogy in generating engagement in the learning process. Finally, we present an account of how the language used in analogical reasoning offers us both possibility and constraint in shaping the way we conceptualise the world.

The role of analogy in learning has been extensively researched in science education. The core purpose of the use of analogy as a strategy deployed in teaching is that of developing understanding of abstract phenomena from concrete reference. While such an objective is desirable, it is based on the assumption that there is an agreed interpretation of the particular phenomena under scrutiny to which all subscribe. This is not without contention, and we will later explore the pedagogical implications of adopting such a position.

The use of analogy is both a pedagogic strategy deployed in science education to support understanding of phenomena and a cognitive tool for the development of theoretical insight in scientific enquiry. Dreistadt (1968) catalogues the central role of analogies in the history of scientific ideas including the works of Einstein, Darwin, Bohr, Mendeleev and Kekule. The importance of analogy in the history of ideas in science is further exemplified by Treagust (2007) in reference to the seventeenth-century astronomer Johannes Kepler. That analogy is employed in science reasoning is not entirely a surprise since explanation of the abstract needs to be rooted in existing experience in order to interpret increasingly sophisticated ideas. In reasoning through analogy, the scientist in the scientific endeavour and the student and teacher in science education are involved in the same process. Developing meaningful explanation could, therefore, be considered the core enterprise of both the scientific endeavour as well as personal learning in science. That this is the principal task in pedagogy is supported by the plethora of research literature focused on learners' misconceptions across a wide range of conceptual domains (Duit 2008).

D. Heywood and J. Parker, *The Pedagogy of Physical Science*, Contemporary Trends and Issues in Science Education, vol. 38, DOI 10.1007/978-1-4020-5271-2_3,

3.1 Learning About Simple Circuits

The National Curriculum for Science in England (DfEE/QCA1999) demands that primary school children should be taught to construct simple circuits (including series circuits)and be taught how to vary current in a circuit, and that they should develop the notion of a complete circuit during their education. Although the curriculum does not, explicitly, entail developing explanations for the phenomena observed, this does not mean that children do not possess and utilise their own explanations and reasoning to make sense of their observations (see for example, Osborne and Freyberg 1985; Osborne et al. 1991). Since the development of explanation is core to learning and teaching in science, both teachers and children when engaged in scientific activity will, inevitably, be concerned with building qualitative understanding of the behaviour of simple circuits.

It does not seem adequate, in terms of providing teachers with the confidence to engage effectively with children's ideas on simple circuits, to simply address practical issues of circuit building, which, arguably, is essentially a technological enterprise anyway. In any event, children will have ideas and raise questions about how electric circuits function: what makes a bulb light, why are two bulbs less bright than one in a series circuit, or what causes the battery to go flat? Easley (1990) found that children's questions were often focused on the mechanisms of how things work and that standard textbooks provided little by way of supporting qualitative understanding in teaching and learning. This presents a significant pedagogical challenge to teachers in developing appropriate explanations for learners who want to know, at a qualitative level, what is happening to make a bulb light in a simple electric circuit. A common approach to learning in this area is through the use of analogy, attempting to make more accessible to learners what is essentially abstract and intangible because it cannot itself be seen: only manifestations of 'it' are evident (for example, a bulb lighting).

Analogies, in a general sense, have been researched by Clement (1993) in a study on high school students' preconceptions in physics, and Wong (1993) on trainee teachers and their use of their own analogies in generating explanations of physical phenomena. As mentioned above, the central role of analogies in the history of science parallels that of their deployment in pedagogy. Analogies are central to reasoning in science in that they show how *parts* are put together into *wholes* or how to differentiate the *parts* of *wholes*, so that problems are solved. It is possible to deploy their use in analytical reasoning to illustrate the nature of the parts and their relation to the whole. An example of this kind of reasoning is Rutherford's solar system model of the structure of the atom. In such reasoning, it is essential to be able to transfer from the analogy those elements that are most productive in promoting understanding of the problem under investigation. Thus, the thinking involves the constant juxtaposition from what Gentner and Gentner (1983) term the base domain (the analogy) to the target domain (the concept). Differentiating those parts that are useful from those that clearly are not relevant helps promote further understanding. An example of this would be the attributes of the solar system that provide a 'picture' of the model of the atom (for example, the Sun as the central nucleus and the planets as peripheral electrons moving around) and those that would not be relevant (for example, the sun's temperature or the colour of the planets).

How effective analogies are in promoting conceptual understanding with respect to electricity has been researched with children of varying age (Gentner and Gentner 1983; Shipstone 1984; Tiberghein 1985; Cosgrove 1995). Building mental imagery from parts to explain the whole can provide a powerful tool for overcoming a common difficulty in understanding the behaviour of electric circuits, that of conservation of current. The findings of Gentner and Gentner indicate that some analogies are more successful than others in developing reasoning about a specific idea and the comparison of a 'moving crowds' model with a water model for electric circuits provided evidence that the former was more effective in developing understanding of this particular concept. No single analogy will promote total understanding of the complexities of electricity.

This raises the issue of how best to support teachers in their learning to promote their qualitative understanding of what is happening in electric circuits, that is, to focus on learning about, rather than just 'doing' electricity. Understanding circuits involves dealing with a set of interrelated concepts such as current conservation, resistance and energy transfer. These are often dealt with separately through analogies that seem to inadvertently promote a sequential rather than a holistic view, and this causes particular problems in learning, the implications of which will be discussed later. It would seem sensible, therefore, to focus on the generative power that such intellectual engagement fosters rather than search for an all-embracing analogy. Such reasoning should not be avoided simply because it raises doubt in the learner's mind, but should be used productively to support teachers in recognising that this is part of the nature of science such that they will have expertise in recognising similar processes in their pupils' learning and have the confidence to encourage them to pursue such activity. Indeed, the curriculum in England places strong emphasis on the generation of explanation in science learning in stipulating that pupils should be taught that science is about thinking creatively to try to explain how living and non-living things work, and to establish links between causes and effects. (http://curriculum.qca.org.uk accessed July 2008).

Developing insight into the powerful but provisional nature of scientific knowledge and explanation will raise pupils' awareness of the process by which scientific models are created, tested and modified. Most importantly, it affords the opportunity to develop a sense of the excitement of scientific discovery that fuels the unceasing inspiration of scientists.

The case study reported here explores the use and limitations of analogy in supporting experienced primary teachers' learning about electricity on a course specifically designed for the purpose of developing their knowledge and understanding in order to enable them to teach the science curriculum effectively and with confidence. Although the course was principally centred on providing such support, it addressed wider pedagogic concerns to raise teachers' awareness of issues in learning science to broaden their professional repertoire. This included reflecting on their knowledge and understanding about the nature of the scientific process as experienced in their own learning on the course, and its relevance in promoting effective pedagogy. As Cosgrove (1995) illustrates, learning in science is not simply a case of teachers imparting scientific knowledge to pupils rather, it involves encouraging the learner to generate their own ideas through intellectual scrutiny of the problem. In order to do this effectively, the teacher needs a command of the conceptual issues in electricity to engage confidently with the children, to encourage their questioning and help 'construct' their ideas about what is happening. This requires a considerable pedagogic awareness and sensitivity. It is our contention that it is a fruitless crusade to search for the Holy Grail of analogies that will satisfy all concerns and questions for the learner. No such analogy exists, and nor could it, because analogies are attempts to relate abstract phenomena to existing experience and concepts. The focus of the study was to explore the usefulness of typical analogies used in the teaching of electricity, with reference to the learning of primary teachers, whilst providing the opportunity to explore their usefulness and limitations through practical investigations and critical scrutiny.

The research was carried out during one session provided for 25 in-service primary teachers (subsequently referred to as students in the reporting of the findings), undertaking a short course designed to improve their subject knowledge in science (see Heywood and Parker 1997 for further details). The focus for the session was to develop understanding of simple electric circuits through the use of analogies. Students worked in small groups investigating and discussing their understanding of simple electrical circuits. Initially, each group was asked to light a bulb using a battery and some wire and to collect together their ideas about what they thought was happening in the circuit. Finally, they listed any questions emerging from their observations and this enabled the establishment of the general nature of starting points. As the students explored simple electrical circuits they were introduced to a range of typical analogies and, at strategic points, they were asked to write individual reflections on how useful they had found these analogies to be in their own thinking and learning. They were required to qualify their comments by being specific about the ideas that had been developed as a result. A group record was kept of emergent questions throughout the session. In addition to individual reflection and the collection of group ideas and questions, tape recordings were made of group discussions and students were interviewed in order to provide detailed comment on how the analogies had influenced their learning in this area.

3.2 Applying Analogies to Simple Circuits

3.2.1 Analogies Deployed

3.2.1.1 Analogy 1

The human circulatory system was used as a simple analogy of a *closed* system in that the heart acts as a pump in order to pump blood around the body and the amount of blood leaving the heart is ultimately the same as the amount of blood entering the heart. However, the nature of the blood, with respect to oxygen level has changed. It was anticipated that students may develop/consolidate certain ideas and these might include:

Base domain	Target domain
Constant blood volume	Current conservation
Heart acts as pump	Function of battery
Oxygenation level of blood	Changed nature of electricity

3.2.1.2 Analogy 2

This analogy was introduced through a role-play activity in which individuals represented the movement of *electrons* in a circuit. The notion of electrons/electricity was not explored at depth in a scientific sense; rather the terminology was adopted from the way in which students tended to describe flow in an electrical circuit. They walked at constant speed around a circle and, in doing so, passed through a *battery* that was represented by one member acting as an energising point and causing individuals to wave their arms wildly as they walked. Within the circle the individuals were required to pass through a narrow tunnel where their arm movements were restricted resulting in arms frequently hitting the walls of the tunnel making it *heat up* (*bulb glows*). Individuals walked back to the battery where they were re-energized before completing the circuit again. This analogy produced much debate, and it was possible to identify several possible ideas that might be developed/consolidated as a result. Examples are listed below:

Base domain	Target domain
Individuals walking	Current
Individuals returning to battery	Current conservation
Speed of walking	Current size
Restriction of arm movement in tunnel	Energy transfer
Tunnel becoming hot	Energy transfer resulting in bulb glowing
Size of tunnel	Resistance in circuit
Cause of arm movement	Role of battery
Magnitude of arm movement	Size of battery

3.2.1.3 Analogy 3

This is a typical analogy employed in the teaching of electricity that involves the notion of a closed water system in which a pump circulates water and into which constrictions can be introduced. Such constrictions will have an impact on pressure in the whole system.

Base domain	Target domain
Water flow	Current
Pressure in the system	Response of whole circuit
Constrictions	Resistance in circuit
Pump	Battery
Relationship between constriction, pressure and flow	Relationship between resistance, voltage and current

3.2.2 Synopsis of Research Findings

3.2.2.1 Initial Ideas About a Simple Circuit (Fig. 3.1)

Figure 3.1 indicates the range of ideas expressed about a simple circuit with one bulb. The ideas have been presented in categories concerned with the circuit, the flow, the battery and the bulb. All group diagrams included a complete circuit; however, only three groups stated explicitly that a complete circuit was needed in order to make the bulb light. Initial ideas across the groups indicated a wide range; with Group A having a limited insight compared with groups E, D and F.

The groups' initial questions (Fig. 3.2) ranged from general considerations to specific questions focusing on the flow of electricity and the nature of the circuit, what was happening at the bulb and the function of the battery.

Groups	**A**	**B**	**C**	**D**	**E**	**F**
The circuit:-						
complete circuit drawn	√	√	√	√	√	√
states complete circuit needed	√				√	√
unidirectional flow			√		√	√
unsure of direction flow	√	√	√	√		
The flow:-						
something flows			√			
electricity flows		√			√	√
energy flows				√		
electrons carry electricity					√	
something returns				√		
electricity returns						√
nothing returns		√				
unsure about what returns		√	√			
The battery:-						
questions function	√					
stores *something*			√			
stores power		√				
stores energy		√		√	√	
The bulb:-						
something happens at the bulb				√		√
something makes the bulb light		√				
something is used up				√		
energy is used up			√		√	
electrons are used up						√
energy is transferred/converted					√	
something turns into heat & light				√		
energy turns into heat & light				√	√	√

Fig. 3.1 Initial ideas about a simple circuit (N = 25)

1. General questions: -

What is happening?
What is happening at the bulb?
What is happening at the battery?
What is going around the circuit?
Why do you need a complete circuit?

2. About the bulb: -

Why won't the bulb light with just one end connected?
Why does the wire in the bulb light and the others don't?
What is it that makes the bulb light up?
Why is it turning heat into light

3. About the flow of electricity: -

Which way does the electricity flow?
Why does it flow from +ve to –ve?
Does something flow back from the bulb?
Does some electricity flow back to the battery?
Is electricity coming out of both ends of the battery?
Why can't it come out of both ends of the battery?
If stored energy is going from the battery to the bulb then why do you need a return wire?
Does the position of the wires matter?

4. About the battery: -

Why does the battery go flat?
What's inside the battery?
What's the difference between the positive and negative ends?
What is the power in the battery?

5. Miscellaneous: -

Why metal? How does it get through the plastic wire?
What is a good conductor?
What does a.c. mean?
What is an electron and how does it work?

Fig. 3.2 Initial questions

3.2.2.2 Analogy 1 (Fig. 3.3)

Twenty-four out of 25 students found Analogy 1 to be useful in their learning. The main reason given was with respect to the concept of current conservation. A limited number of questions evolved from this analogy with only two students reasoning that if current is conserved then *something* must be *lost* to cause the battery to *run out.* One student expressed difficulty with the base domain of blood and circulation.

3.2.2.3 Analogy 2 (Fig. 3.3)

Twenty-four out of 25 students found Analogy 2 useful. Fifteen reported that the analogy helped to develop their ideas about current conservation, 16 identified that

Usefulness: -	**Analogy 1**	**Analogy 2**	**Analogy 3**
useful in developing thinking	24 (96%)	24 (96%)	16 (64%)
not useful in developing thinking	1 (4%)	2 (8%)	8 (36%)
Reasons provided: -			
current conservation	20 (80%)	15 (60%)	8 (36%)
changed state of electricity returning	12 (48%)	-	-
energy transfer at bulb	-	16 (64%)	-
role of battery	-	10 (40%)	-
resistance in the circuit	-	-	5 (20%)
whole view of the circuit	-	-	4 (16%)
The analogy: -			
provides a concrete example	5 (20%)	6 (24%)	-
made sense/was readily understood	3 (12%)	-	3 (12%)
reinforced previous ideas	1 (4%)	3 (12%)	-
was more effective than previous analogy	-	3 (12%)	4 (16%)
difficult to visualise	-	-	1 (4%)
difficult to transfer thinking from one analogy to another	-	-	2 (8%)

Fig. 3.3 Usefulness of analogies (N = 25, percentages in brackets)

the analogy explained the idea of energy transfer at the bulb and ten students found it explained more clearly the role of the battery.

On critical scrutiny of the second analogy some students recognised that it failed to explain whether electrons lose all or some of their energy at the bulb and also the difference between energy transfer at the bulb and energy needed for electrons to move around the circuit. The failure of the second analogy to explain circuit behaviour generated a much greater range and depth of questioning than Analogy 1 where translation was considered relatively unproblematic in terms of the concept of current conservation. Analogy 2 resulted in ten questions, compared to only four in analogy 1.

3.2.2.4 Applying Analogy 2 to Two Bulbs Wired in Series

On being asked to apply the second analogy to a two-bulb series circuit, 22 students recognised that the analogy had broken down. Significantly, 15 students identified that the analogy would still hold if a whole view of the circuit was adopted. Two students recognised that explanations were not consistent with a sequential model of a circuit which was the principal idea generated by the analogy under consideration.

There was a significant increase in the generation of questions when the analogy broke down; 19 of the students' questions focused on the relationship between the two bulbs with respect to the circuit as a whole including concerns about how the

Analogy fails to explain observations	22 (88%)
Analogy works if *whole* circuit view adopted	15 (60%)
Analogy not consistent with sequential mode	12 (8%)
Analogy breakdown useful in promoting thinking	3 (12%)
Analogy breakdown confusing/frustrating	5 (20%)
Analogy breakdown prevents learning transfer	1 (4%)
Questions arising:-	
How does the circuit *know* energy has to be shared?	5 (20%)
Why isn't all the energy used up at the first bulb?	4 (16%)
Why does the second bulb light at all?	1 (4%)
Does the first bulb *know* about the second bulb?	1 (4%)
Why do the bulbs glow dimmer?	2 (8%)
Why isn't the second bulb brighter?	2 (8%)
If there was less wire would the second bulb be brighter?	2 (8%)
Why doesn't the position of the bulbs in the circuit matter?	2 (8%)
Does the circuit behave as a whole or a sequence?	1 (4%)
Are the bulbs making a larger tunnel?	2 (8%)
Does the position of the battery matter?	1 (4%)
How is the battery capable of giving out more energy?	1 (4%)
Does the battery continuously lose power as it lights the bulb?	1 (4%)
Would an ammeter show a difference in current here?	1 (4%)
What is the relation between watts and volts?	1 (4%)

Fig. 3.4 Outcomes of applying Analogy 2 to two bulbs in series (N=25, percentages given in brackets)

circuit *knows* that energy has to be shared (15 students) and why all the energy is not *used up* at the first bulb (four students). In total, 27 questions were raised as a result of Analogy 2 breaking down in the light of practical investigation compared with only four questions raised by Analogy 1 in which the students appeared comfortable with the effectiveness of the analogy in explaining current conservation.

3.2.2.5 Analogy 3 (Fig. 3.3)

Sixteen students found the analogy useful, whilst eight found it failed to help their thinking, and one person reported it as being neither helpful nor unhelpful. One student had found difficulty in visualising the base domain concept of *water pressure/water flow*. The analogy was found to be useful with respect to promoting/reinforcing current conservation, conceptualising resistance in a circuit, developing a holistic view of a circuit and in increasing student confidence. Four students found Analogy 3 more effective than previous analogies with two students expressing difficulty in transferring their thinking from one analogy to another, and a further two students experiencing some difficulty in visualising the analogy, finding it too *abstract*.

3.2.3 *Tracking Learning Within the Groups*

The following more detailed analysis of results has been presented in order to illustrate the extremes of position across the different groups.

3.2.3.1 Group A

The group's initial ideas about the simple circuit were extremely limited. They knew, for example, that a complete circuit was needed in order to light the bulb but were unsure as to why this was necessary and *what* was flowing in the circuit. Initially, the group generated the least number of questions of all the groups and their questions were not directly concerned with qualitative notions of what was happening in the circuit.

After the first analogy, which the group found useful in explaining current conservation, their concerns became centred on *what* was being *lost/used up* in the circuit. They reported finding the second analogy useful with respect to developing notions of current conservation, energy transfer and the role of the battery, and moved on to ask much more detailed questions about what happens when electrons travel through a bulb, how much energy is *lost* and why the battery *runs out* of energy.

On applying Analogy 2 to a two-bulb series circuit, there was recognition that the analogy no longer explained the phenomenon adequately. They expressed their difficulty in reconciling the evidence with the analogy and became concerned about issues such as *how does the circuit know that energy has to be shared?* Analogy 3 encouraged the group to adopt a whole view of the circuit in order to explain their findings.

3.2.3.2 Group D

The group's original ideas were reasonably well-developed; for example, they indicated the notion that the battery was a store of chemical energy and that there was a flow from this battery to the bulb that resulted in the heating and glowing of the filament. Concerns centred on what was happening at the bulb and why the wire was glowing when other wires in the circuit were not. The direction of current flow and how the positioning of the parts might affect outcomes were also questioned.

Analogy 1 was found useful with respect to current conservation and the *changed state* of electricity as it passed through the bulb. A singular reservation was expressed by one student who reported that it was possible to visualise oxygen but more difficult to visualise energy. One student found the analogy helpful in explaining the role of the battery. The group began to think more deeply about what was happening at the bulb and raised questions about the possibility of there being two *types* of energy (energy *for walking and energy for waving*).

On application of Analogy 2 to the two-bulb series circuit, all the group members reported that the analogy had broken down as an explanation. As a result of this two of the five members considered this had stimulated their thinking and three of the group found it to be a frustrating experience. Concerns focused on the need to resolve the conceptual conflict through obtaining definitive answers to questions about observations of circuit behaviour which could not be explained through application of the analogy, for example, *why was there reduced brightness of the bulbs, why did this not happen with bulbs in parallel, how was the energy from the battery being controlled?* Analogy 3 was helpful to three out of the five students but not to the others. In general they tended to offer no explanations for their comments.

3.2.3.3 Group C

Initially, the group drew a complete circuit and indicated that they thought that the battery was a store of energy and that most of the energy was *used up* at the bulb resulting in heat and light. They indicated unidirectional flow in the circuit (−ve to +ve) although their questions indicated that they were unsure of the direction. They were also unsure as to why a return wire was needed as, according to their model, the store of energy was transferred at the bulb and turned into heat and light. Other questions included why metal is used in the making of wire and how the energy gets through the plastic of the wire.

All members of the group found Analogy 1 helpful in developing ideas on current conservation and this resulted in critical questioning. Having used an ammeter in the circuit in order to confirm that the current was indeed conserved, one student commented:

> [*the analogy*] ... leads you to think that the blood is carrying something around the body, oxygen ... gaining the use of that oxygen, so there seems to be a loss there and there wasn't a loss in the electrical circuit.

On the introduction of Analogy 2, the group again reported this as being helpful with respect to the concept of current conservation and in generating ideas on energy transfer at the bulb and the role of the battery. They moved on to wrestle with their original concern of what happens at the bulb. Their reasoning centred on the attributes and processes within the base domain. They began to talk of two types of energy (*walking energy and waving energy*) and wanted to know if energy was left on leaving the tunnel and one member was unsure as to how the battery was able to provide the *kick* in the circuit.

On application of Analogy 2 to the two-bulb series circuit, all members reported that the analogy broke down as an adequate explanation for their observations. Significantly, they were able to recognise that the reasons for this were associated with their attempt to apply a sequential model to the behaviour of the circuit:

> ….if the electricity is going to the first bulb ... that should take as much as needed last time when we only had one bulb ... but this bulb doesn't know that this bulb is in the circuit.

Another student commented when comparing the relative brightness of bulbs that:

> .....if energy is coming out of here and jiggling along the wire I'd think it was the first bulb that would have the energy and then it would be reduced for the second bulb and coming back into the battery for another kick.

On testing this by swapping the bulbs, the students observed that this did not fit with the evidence:

> .....it's not reversed because when that was over here that was the brightest bulb ... so it's not the position that's the problem.

They began to explore the notion of a whole circuit:

> .....as soon as you plug the battery in the whole thing moves ... it doesn't start here ... everything else from here doesn't move ... I'm proposing that this battery causes all the little whatevers ... the electrons ... to move simultaneously.

On the introduction of Analogy 3, half the group found this to be a useful and satisfying explanation, while others still struggled to come to terms with the model, the persistence of the sequential view proving intransigent:

> .....surely any analogy with one energising point must be sequential.

They subsequently reported that a whole circuit view *describes* what happens but does not *explain* it.

3.3 Implications for Pedagogy

The research supports the contention that the usefulness of an analogy seems largely dependent on whether the base domain resonates with the learner's existing experience. Where it does not, the learner is unlikely to be able to transfer to the target domain. In any event, because learners have a wide range of experience and bring different agendas to the learning situation, the purposes of analogies are perceived differently. We believe that individual perceptions are dependent on the learner's past experience, current explanations and beliefs of what constitutes learning in science. Our findings indicate that when learners with established ideas about the workings of a circuit were challenged, there was a reluctance to generate questions. The converse was true of those learners who had less formalised ideas on electricity. The evolution of questions in the latter case was greater in both quantity and depth.

3.3.1 The Problem of Analogies in Developing a Sequential View of Simple Circuits

Using analogies in differentiating the parts of the whole is not without problems in understanding the nature of how circuits behave. This is particularly apparent in relation to difficulties encountered in applying a sequential view to the behaviour of circuits. All groups experienced difficulty in resolving this problem and some were able to identify this explicitly as being a stumbling block to their own learning. This seems to be a conceptual hurdle at which most people fall and is supported by the findings of Shipstone (1984) who found that 80% of 13-year olds hold this view. As Driver (1994: 120) indicates:

> It is notable that prevalent alternative models are sequential models in which something from the battery travels around the circuit, wires and components in sequence. This deep-seated notion ... underlies many of the problems which pupils have in understanding the behaviour of electric circuits. It is the notion that might have been considered as the underlying mental model having various expressions.

We believe that analogies such as the moving crowd model actively promote a sequential view of a circuit and that this seems particularly intransigent to change when challenged. In this sense, the analogy is 'hoist by its own petard' in that it resonates so closely with the learners existing experience, that transfer from the base domain results in such strong conceptual coherence that learners are reluctant to challenge apparent limitations. Consequently, the part has obscured the whole or has become the whole.

The significance of this in our findings is that in such circumstances, the questioning process may stop and because the learner is comfortable with what is learned, he/she is reluctant (or unable) to perceive the need for further engagement. The incidence of questions is reduced to the extent that the learner is unable to perceive how the model could be put to the test of empirical evidence which would either support or refute predictions of the behaviour of electric circuits based on the analogy (for example, two bulbs of different brightness in a series circuit). Being *comfortable* satisfies the desire for challenging engagement. This has implications for teachers' perceptions as to the nature of scientific endeavour. If learning in science is viewed simplistically as a matter of providing illustrations for the effective transmission of explanations, then the central task in this case would be to find the *right* analogy to resolve all conceptual anxiety. Such pursuit of the perfect analogies to explain complex ideas is likely to prove fruitless. As one student commented when confronting the limitations of analogies in explaining the behaviour of two bulbs in a series circuit:

> I don't want an analogy now ... I want the facts. I don't know what the facts are, I want someone to tell me ... every time we get an analogy they come to an end and there's always something you need an answer to.

The frustration expressed here underpins the notion that there are facts that can provide a mechanistic explanation for observations. As Feynman (1992: 37) pointed out with respect to Newton's Law of Gravitation:

> Newton was originally asked about his theory 'But it doesn't mean anything – it doesn't tell us anything'. He said, 'It tells you *how* it moves. That should be enough. I have told you how it moves, not why'. But people often are unsatisfied without a mechanism.

Thus, the pedagogic challenge may well be to ask the question: Is it possible to build *bridges* for learners in such circumstances by using a sequence of analogies that facilitate the learner's journey from base to target domain such that the inherent problems of sequential notions of a circuit become less persistent? This adds a different perspective to the 'parts to whole' (whole to parts) approach. Browne (1994) has investigated the merits of such techniques in an interesting study on pupils' ideas on forces. He explores facilitating conceptual change through presenting *bridges* between the base and target domain using analogies that progressively build on existing experience towards more abstract notions to explain target domain behaviour. Such axiomatic reasoning is an approach designed to make the abstract more tangible by relating the concept broadly to already established, or at least believable, ideas. We argue that the extent to which such a process is transferable to increasingly complex concepts which are interrelated (for example, current conservation, resistance, energy transfer) is questionable. Our findings indicate that teachers found considerable difficulty in making the conceptual journey from a sequential to a holistic view. Where they were comfortable with an analogy that proved a satisfactory explanation for a *part* of a *whole* (in, for example, current conservation), they were unable to transfer from this sequential perspective. This is illustrated in the following student's comment after the third analogy that had followed considerable engagement with explaining the behaviour of circuits through different analogies; a literal exposition of breaking the whole into parts:

> .....surely any analogy with one energising point must be sequential.

This has implications for a pedagogy based on constructing meaning for others through explanations of the abstract through analogy. However, in our opinion, the essential element of this sort of linear progression of reasoning is not that the learner is comfortable, or has achieved the *right* scientific viewpoint through using the analogy; rather, the issue concerns the quality and the incidence of self-questioning the learner generates when an analogy breaks down under critical scrutiny. It is only through such intellectual endeavour that the complex nature of abstract ideas is eventually understood. The study provides evidence that combining, building on, moving between analogies and rigorous examination of ideas through practical activity enhances learning.

It is in the area of scientific reasoning that the optimum opportunity arises, when an analogy breaks down, to focus the student's attention to the fact that such experience is a core element of the scientific enterprise. This has implications for teaching style, in particular, how teachers use their knowledge and what knowledge teachers need (i.e. knowledge of both the concept and the nature of scientific reasoning) in order to feel confident to use such approaches effectively. Concept acquisition and scientific reasoning are not mutually exclusive in this process. Conceptual understanding can continue to develop when an analogy no longer supports the learner in making links from the base domain to the target domain and it is our premise that when this occurs, learning can be enhanced through the questioning process. Such increase in the depth and incidence of questioning is evident in our findings when analogy no longer fits with empirical evidence. The questioning increased from 4 (when the teachers felt comfortable and were convinced of the usefulness of the first analogy in explaining current conservation) to 27 when the analogy became incoherent on being applied to two bulbs in a series circuit.

The importance of this should not be underestimated as it leads the learner to consider and search for alternative explanations using a multiplicity of analogies to generate further understanding. Other analogies would, in such circumstances, either evolve from the learner or be provided by the teacher. In any event, the limitations of all analogies in learning would be realised and a simplistic notion of concept transfer would be exposed as false pedagogy. This is fundamental for teachers to recognise.

In a constructivist paradigm in which the learner is actively constructing meaning through analogy, this poses a significant problem. First, in order to use analogies to make bridges progressively from base to target it assumes that there is complete agreement as to the nature of the target concept; this is a highly contentious notion in the subject area of electricity. Second, there is a fundamental problem if the analogy cannot facilitate the bridging of the base and target domains. There is considerable tension between professing to promote critical engagement and handling learners' anxiety when they are unable to make something fit. The task in pedagogy concerns the development of professional insight into handling the subtle nuances in teaching and learning when learners experience such cognitive conflict, an issue that is addressed in more detail in Chapter 6. Research papers on the role of analogy in promoting understanding of electricity are conspicuous in their lack of qualitative explanation of what the target domain (electricity) *actually is.* There is an almost implicit assumption that there is complete agreement by science educators as to the nature of the phenomenon. Even a cursory conceptual exploration and discussion with fellow colleagues is likely to reveal this not to be the case (see for example, Black and Harlen's (1993) account of a physicist's view of electricity). This is not surprising since there is no analogy or combination of analogies that can fully explain the empirical evidence observed in even the simplest circuits. Where they do appear to do so, it is often because the analogy has not been subject to critical scrutiny by the learner.

Thus, the essence of our argument is that it is the promotion of the critical scrutiny in challenging the analogy, attempting to apply it, recognising when and why it breaks down that the opportunity for learning really takes place. The focus is, in this sense, diverted from attempts to find the perfect analogy towards accepting that there are limits to understanding, and that we need to encourage such intellectual engagement in order to promote effective teaching and learning in science. That is, to encourage learners to recognise that this is part of the scientific enterprise and to focus on the extent to which their understanding has developed.

Such questioning processes between the learner and themselves or fellow students and teacher and learner during self-generated analogy are what Cosgrove (1995) terms *science-in-the-making*. Despite his finding the persistent problem of the sequential view of an electric circuit being evident in the students' ideas, he also found that students recognised the limits of their own analogies. They demonstrated the capacity to use this positively in addressing the limitation of their own and other analogies (including the teacher's explanations) to develop thinking further. In short, when the analogy breaks down, then you have science in the making. Interestingly, the findings of Cosgrove revealed that after a considerable period of the syllabus, not one of the students knew what their teachers' understanding of electric circuits was. This is not to suggest that the teacher did not have a view; rather, it places a different emphasis on both what the teacher needs knowledge of and how such knowledge is used. It is unlikely that individual understanding could be interpreted directly (by the learner from the teacher) no matter how effective the constructs of parts to whole, no matter how seductive an analogy appears to be at a particular point in time. The interpretation is always partial. We are constantly building concepts, inventing stories (analogies) and questioning whether those analogies hold up under scrutiny in order to move our own understanding that little bit further on. The ground is forever shifting. This is the essence of the symbiotic nature of teaching and learning. We would argue that entering into honest discourse with learners as to the difficulties we all encounter when wrestling with such ideas would promote a more positive view of the nature of learning in science.

3.4 Explanation and Meaning

In Section 3.5 we examine some of the pedagogical implications of the research findings particularly in relation to explanation and the language used to access, through analogy, that which is not only counterintuitive but also intangible. In this discussion, the pedagogic task is conceptualised as being concerned with what Eger (1992a) considers the central issue of science education; the problem of meaning. The complexity of the pedagogic challenge in this regard concerns the unnatural nature of scientific explanations of the world that conflicts with life-world experience (Wolpert 1992). Intuitive notions of phenomena such as forces, energy, seasons and electric circuits are seldom congruent with scientific explanation. Science seems incompatible with life-world experience. These counterintuitive scientific explanations constitute what (Eger, 1992a, b; 1993) refers to as *the text* of science. The use of the term needs some qualification. What is being referred to here and subsequently as *text* is that which both teacher and student encounter in science education discourse, the parameters of which are traditionally considered as being defined by textbooks, curriculum syllabi, and to a lesser extent, in school science at least, science education research. This is different from the classical metaphor of science research as a 'reading of the book of nature' in which the scientist is involved in deploying scientific method to determine how the physical and living world works.

The problem that both teacher and student face in regard to this text is that of interpretation leading to scientific explanations that are underpinned with causal mechanisms that are meaningful and coherent. This could be considered the underlying rationale for the teaching through analogy of abstract phenomena in order that the learner can develop causal mechanisms to explain the behaviour of simple circuits. However, within this enterprise, there is implicit the notion that science text is somehow given a priori and stands independent of discourse in science learning and teaching. The conspicuous absence as to what constitutes the scientifically acceptable in the misconceptions literature in science education seems to underline this view. In the proceeding discussion we shall argue that such an epistemological position requires further scrutiny.

3.4.1 The Appropriation of Hermeneutics

It is in regard to the focus on meaning from the interpretation of text in science learning and teaching that the appropriation of the discipline hermeneutics emerges, the modern origins of which evolved from the interpretation of ancient texts at the turn of the nineteenth century. This has been illustrated for example, through the work of Gallagher (1992a, b) who has developed a comprehensive account of the application of hermeneutic principles to education and the work of Eger (1992a, b, 1993). Eger provides a detailed examination of the philosophical issues concerning the appropriation of hermeneutic inquiry to both *science* and *science education*, the latter of which is the central concern of our discussion here. The practical implications of this work for the development of teacher subject and pedagogical knowledge in science has also been explored (Heywood 1999), and the potential for further inquiry, whilst contentious, does offer the opportunity to creatively engage with some of the *taken for granted* assumptions on which some science education research is based. Whist it is not possible to deal with the principles of hermeneutics in depth (we explore this in more detail in Chapter 5), for the purposes of this discussion we shall attempt to underline its relevance in terms of the nature of scientific explanation and, more importantly, how these explanations are presented in science education.

In hermeneutic inquiry the absolute meaning of a text is deferred because there always remains the possibility of alternative interpretation. This notion of *undecidability* sits uneasily with a conceptualisation of science as knowledge of how things are, in which it is argued that competing theories are *decided* upon the basis of empirical evidence; an often misrepresented notion of conjecture and refutation. The mechanisms underpinning the 'method' of inquiry in respect of this have been the central focus of twentieth-century debate in the philosophy of science (see for example Kuhn 1970; Popper 1963). However, we would contend that the issue of *decidability* has had a pervasive influence on the presentation of science knowledge in the curriculum such that *knowing* is privileged over *understanding*, *practical inquiry* over *discourse* and, *method* over *interpretation*.

3.4.2 Exemplification of Language and Meaning

There arises from this a significant issue that is seldom recognised in science education research literature on subject and pedagogic knowledge concerning the perception as to where meaning resides. Hermeneutics challenges the notion that *meanings* await expression independent of the language used to describe them (Brown 1997), and we offer two examples to illustrate the point. First, take the idea of fields that evolved from the work of Faraday on electricity and magnetism, a comprehensive historical treatment of which is given by Gooding (1989). Faraday initially developed the concept from meticulous experimental procedures including the observed effect of magnets on iron filing patterns. Subsequently, these effects were eventually interpreted as evidence for the existence of a field of force exerting influence on matter. The field description (i.e. the language of fields) that had derived from observed magnetic effects then evolved an independent ontology through Faraday's interpreted *reconstruction*.

> The process of reconstruction makes natural phenomena – which have been accessible only through human construction and intervention – come to be accepted as independent of that activity. The residue of phenomena thus came to appear as objects independent of human intervention. (Gooding 1989:216) [author emphasis]

The reading of natural phenomena then comes about through the human act of interpretation. Historically, the idea of fields developed into a generic concept useful in describing a diverse range of physical phenomena (the electromagnetic field). The abstraction has since become increasingly remote from concrete models and analogous reasoning, such that the epistemology of science as an ontological reading of how things are is derived from abstracted interpretation *in language*. In learning science, this requires an integration of interpretation, the difficulty of which should not be underestimated. The physicist Feynman (1992) acknowledged that conceptualising the abstract notion of fields in concrete analogous terms was a pointless endeavour because no analogous equivalent exists.

As a second example, consider the abstraction of ideas concerning waves derived from the concrete experience analogous with water. This paradigm has progressed from the concrete visualisation that parallels wave behaviour in a recognisable medium (in this case water) towards abstract mathematical models. Along the course of this historical development of wave theory, the medium through which electromagnetic waves travel (*the ether*) was discarded, leaving only the waves themselves, a metaphysical position which has no analogous equivalent. In this there emerges an apparently absurd notion, that of describing the behaviour of waves in a pond whilst maintaining at the same time that the pond has no water. However, it is these descriptions in language through which the phenomena take on meaning; an ontological landscape defined through language.

> As far as we are concerned, ontology recapitulates taxonomy – the way we divide the world in language tells us how we think the world is really put together. (Gregory 1988: 174)

Such an ontological position leads to a situation in which the abstract phenomenon (waves), derived from the concrete entity (water), takes on *meaning* (through the language used to describe it) to become an entity as real as the water itself. The pedagogical question that emerges concerns addressing the implications for science learning within this language paradigm in which there is a shift in emphasis so that we are less concerned with ontological entities (atoms, electrons) than the metaphors, analogies and stories that surround them. Two important principles derive from this. First, *meaning* in science is, quite literally, generated in discourse and does not reside independent of it; it is not *'out there'* awaiting some form of *linguistic expression*. Second, because hermeneutics allows for multiple interpretations, *absolute* meaning is necessarily deferred. This undecidability principle arises because there is always the possibility of 'reading' a text differently. We use analogy to illustrate the appropriation of these principles and identify some of the implications of this for the development of teacher subject and pedagogic knowledge.

3.4.3 Alternative Perspectives on Knowledge Acquisition

The approach to teaching electricity through analogy is an attempt to make more accessible what is essentially abstract and intangible because it cannot itself be experienced. Only manifestations of 'it' are evident (for example, a bulb lighting). The teacher, in explaining electricity through analogy, has made a certain decision as to the most effective way to proceed based on pedagogical experience in which analogical reasoning supports learners in critical engagement with abstract phenomena. However, this locates the object's meaning outside the subject's interpretation and implies that the *science itself* is considered a problem only in regard to cognitive transfer, a task conceptualised in terms of pedagogic structures that facilitate acquisition. It may be more productive to consider a different emphasis that explicitly acknowledges that in developing qualitative explanations of simple electric circuits, interpretation is not only partial but also transient and subjective. In this paradigm, analogies would be conceived as mechanisms through which *hypotheses are generated rather than proved* (Wilbers and Duit 2001).

This latter point leads to a consideration as to how analogies are deployed and what science education research has focused on in determining their usefulness in learning. Sfard (1998) provides two metaphors to describe how analogies could be perceived in terms of the pedagogic task of explanation in relation to the 'problem of meaning' referred to earlier. The first is that of *acquisition*. This positions the *raison d'être* of analogy as acquiring commodity (in this case knowledge of how things are), as not being radically different from the privileging of knowing over understanding. The outcome often manifests itself as the satisfaction of inquisition with the subsequent consequence of closing down engagement. Scott et al. (2007: 47) consider *acquisition* to be 'an inappropriate metaphor' to describe the process of learning because it implies a static rather than dynamic. The second metaphor referred to by Sfard (1998), that of *participation*, has a different set of emphases and opens up possibilities in perceiving the purpose of analogy in science learning rather differently (for example, Heywood and Parker 1997; Heywood 2002). More recent work (Lehrer and Schauble 2006) on modelling through analogy and the capacity of this to engender insight into how ideas are generated across a range of science domains supports the contention that it is the process of engagement in critically reviewing such constructs that is the core contribution of analogy in promoting understanding of both science concepts and how science works; or more accurately what constitutes learning in science. We shall return to this point later after a closer look at some traditional approaches aimed at developing qualitative understanding.

Traditionally, analogy has provided a focus of interest in science education and there is considerable research evidence into its use in developing understanding of scientific concepts (see for example, Gentner and Gentner 1983; Shipstone 1984; Tiberghein 1985; Clement 1993; Wong 1993; Cosgrove 1995; Clement 2000; Wilbers and Duit, 2001). In developing qualitative insight into simple electric circuits there is a need to construct a causal mechanism for the concepts of current conservation, energy transfer and resistance.

The first idea to contend with, that of current conservation, is a common conceptual difficulty (Tasker and Osborne 1985) and the reason it is difficult and counter-intuitive is because it is natural to think of something material being used up at the bulb; batteries do not last forever. The idea in *isolation* is relatively easy to address and, as exemplified earlier, is commonly explained through a range of analogies that include closed systems in water circuits, bicycle chains, moving crowds and the circulatory system. In engaging with this idea, the question that presents itself is what exactly is happening at the bulb if current is conserved? In addressing this question, we meet a generic problem in learning: that of conceptualising an abstract idea in relation to other abstract concepts (i.e. current conservation, energy transfer at the bulb and resistance).

The problem is not insurmountable and there is a wide base of literature available that provides commentary on the pragmatics of how this might be achieved. The conservation of current and energy transfer at the bulb as heat and light in a simple circuit is classically explained in analogical terms through variations on a 'moving crowds' model (Gentner and Gentner 1983). That is, people representing electrons moving around a circuit (running track) and meeting a resistance at a certain point (narrow stile through which considerable energy is needed to traverse the obstacle) and energy being transferred at that point. Variations in this idea of energy transfer at the point of resistance include motorcycle couriers depositing 'packets of energy' and lorries in a circular run loading at a depot (battery) and delivering the cargo 'energy' at certain locations (analogous to the resistance) then returning to the depot for reloading.

The pedagogic implications of these and similar approaches have been the subject of considerable debate most of which has focused on two aspects: first, those analogies that are most effective (for example, Summers et al. 1998) and second, the process of how analogies work (Wilbers and Duit, 2001). In the former, as referred to earlier, the usefulness of an analogy seems largely dependent on whether the base domain resonates with existing experience. Where it does not, transfer to the target domain is obscured through the conceptual demand of the base domain. In the latter case, whilst there is reasonable agreement that the learning process demands the constant juxtaposition from the *base domain* (the analogy) to the *target domain* (the phenomena under scrutiny), there is less consensus concerning the finer nuances of the cognitive mechanisms involved. In either focus, however, it is cognitive transfer rather than the science knowledge itself that is considered problematic.

3.4.4 Partitioning and Sequencing

The dilemma in pedagogy centres on how best to address a set of interrelated concepts such as current conservation, resistance and energy transfer at the bulb. Here is exemplified the significant cognitive demand of holding more than one abstract concept simultaneously in relation to others (this is identified as a conceptual challenge in other areas as exemplified in the discussion in Chapter 2 in relation to forces acting on floating and sinking objects). The challenge in explanation concerns the generation of meaning through making accessible complex ideas, and this often involves the strategy of building a holistic conceptualisation from constituent parts. In the teaching of electricity through analogy, these ideas are often discussed separately. However, this differentiating the 'parts'of the 'whole'is not without problems in understanding the nature of how circuits behave specifically with respect to the difficulties encountered with the sequential view of a circuit.

In inadvertently promoting a *sequential* rather than a *holistic* qualitative understanding of the workings of a circuit there is raised the issue of how best to avoid such conceptual confusion in promoting an *acceptable scientific model* of explanation. It has been suggested (Lee and Taw 2001), that conceptualising the circuit in terms of voltage is more appropriate because this enables students to consider the circuit as a holistic system. This view is supported by Frederiksen et al. (1999) who argue persuasively through exemplification of student learning that such an approach is qualitatively coherent in derivation from electrostatic to electrodynamic systems. This introduces the idea of electricity as a flow of charge, but does not necessarily attend to how this can be conceptualised in respect of causal mechanisms that explain energy transfer at the bulb.

Whatever the pedagogic merits and limitations of preferred models of explanation, the most critical issue is the explicit acknowledgement that all analogies breakdown under critical scrutiny, and, therefore, the focus in pedagogy should be less concerned with the search for the Holy Grail of analogies to explain the phenomenon of electricity, than it is with the reasoning that such analogies generate when they break down at critical points in explanation. This has been argued in detail previously (Heywood and Parker 1997), and it is reasonable to question the basis of this contention and examine the implications for the presentation of science knowledge in the curriculum.

3.4.5 The Presentation of Science Knowledge in Science Education

Certainly, $I = V/R$ is a particular way of representing the relationship between current, potential difference and resistance. However, the mathematical expression is particularly limiting in developing a qualitative interpretation of observed phenomenon; there is no causal explanation, for example, of what is happening to make a bulb light in a simple electric circuit. The successive questioning of analogies in relation to empirical evidence would suggest circularity between *explanation* (as offered by the analogy) and *understanding* (of electricity). The analogy offers new ways of understanding observed experience but also governs and informs the way in which it is perceived in future:

> So we have the two arcs of the *hermeneutic circle*, from understanding to explanation (i.e. from understanding the world through a process of categorising it, to making constructions in language in respect of it) and vice versa. (Brown 2001: 36) [our emphasis]

In this, however, there is exemplified the *constraint* and *possibility* of language in analogical reasoning imposes on us, an issue we develop further in Chapter 5. For, while the process could be considered as one of *convergence* in which there is explicit acknowledgement that there is a 'core text' of electricity in existence there to be interpreted, there is no one version as to how things are. In privileging the interpreter, hermeneutics denies that interpretation is simply *reconstructive*; challenging the notion that meaning is situated in the 'text of science' independent of the interpretive subject. From this perspective, the *meaning* that I = V/R has for both learner and teacher could therefore be considered uniquely *individual* rather than uniquely *comprehensible*.

This allows for a range of possibilities and has significant implications for the presentation of science knowledge in the curriculum. It challenges perceptions of learning and the epistemology of science and points towards a radical shift in emphasis of the role of analogy in science learning. In hermeneutic theory, there is a need to recognise the powerful *constraint* and *possibilities* afforded in analogical reasoning. The former is illustrated in the following statement, cited earlier, in which a student when attempting to reconcile the apparent contradictions of a 'tunnel' analogy for resistance in a circuit when applied to two bulbs in series stated that:

> You can only apply the analogy if you are saying the tunnel sometimes becomes larger – you need to view the circuit as a single entity – and the <u>analogy doesn't allow you</u> to do this. [author emphasis]

This exemplification is a potent reminder of the constraint of language in analogical reasoning that in hermeneutics is a necessary condition of interpretation. Without such constraint there would be no recognition that there was something there to interpret. This is illustrated when an analogy is 'hoist by its own petard' because it resonates so closely with the learner's existing experience such that transfer from the base domain to the target domain is considered intrinsically coherent and, engagement with the synthesis of ideas that relate to the phenomenon is closed down. In this condition, the ontological landscape is determined within the base domain and meaning resides *within* the interpretation of the analogy itself rather than *through* the transfer from base to target domain. What is most significant in this is not that the learner is comfortable, or is able to express a scientifically acceptable viewpoint; the issue concerns the *process* of coming to meaning through interpretations that emerge from the quality and incidence of the self-questioning that is generated. This is what is meant by the principle of possibilities afforded in interpretation through analogies. It is in the process of creating this condition, rather than in its explanatory power, that the *potency* of analogy is realised. This has practical implications for the pedagogy outlined below.

3.5 Practical Implications for Pedagogy: Learning

When analogy proves a satisfactory explanation for a 'part' of a 'whole' (in, for example, the concept of current conservation which seems to actively support a sequential model), the holistic interpretation of the circuit is obscured. The importance of model-based learning (Clement 2000) is predicated on the notion that there is a need for science to make sense for the learner and for a satisfactory coherent explanation to be underpinned by causal mechanisms. However, there are limitations with respect to importing from base to target domain and there is a need to monitor the learning process such that those mechanisms that are imported are appropriate and useful. The tension in this concerns the consensus as to what is considered acceptable. In some instances, this is more clearly defined than in others, as for example, in the concept of current conservation. It is when the particular nature of such a 'stand-alone' concept is overlaid not only with other concepts, but in respect of them, that consensus becomes contentious. Science educators are probably in broad agreement that there is a need to consider simple circuits holistically, rather than sequentially, but that a qualitative causal mechanism that underpins an explanation is far less likely to achieve consensus. The implication of this is that the standard 'text of science', when scrutinised beyond surface understanding, is less autonomous in relation to its creation in discourse than is first realised.

This has had significant impact in focusing research attention on the process of conceptual change, with the search for a theoretical model of learning in an effort to identify the factors that influence this process contributing towards the quest for more effective instructional strategies, an issue that is dealt with in more detail in Chapter 2. While learners find it difficult to articulate their ideas and support them through substantive argument (Kelly and Chen 1998), there is potential in this process to explore the transient nature of ideas that are generated through the process of articulation itself. That is, the *act of explanation* necessarily requires scrutiny of the intrinsic coherence of the individual's rationale underpinning existing interpretations.

A further consideration is that of thought experiments. Reiner and Gilbert (2000) recount the importance of thought experiment in the evolution of ideas in science and suggest that the process offers potential for conceptual development and is a frequently used strategy for problem solving. In developing a thought experiment from practical exploration, we are less concerned with the recapitulation of scientifically acceptable ideas established through empirical evidence than we are with the cognitive engagement generated in which understanding is mediated. The thought experiment is in this sense a means of 'testing' the coherence of our own insight into explanations for events observed. The mechanisms of such processes are complex but certainly warrant consideration in the development of ideas in learning. They are *cognitive* rather than *empirical* verifications of hypotheses that again apply equally to teachers and learners. In learning about simple electric circuits through the analogies described earlier, thought experiments about the 'tunnel' being analogous to the bulb in relation to two bulbs in a series circuit revealed anomalies in the intrinsic coherence of the causal mechanisms that were considered to underpin empirical observation. Students readily identified the 'two bulb paradox' manifest in comments relating to how the battery 'knows' to distribute energy transfer evenly in each 'tunnel' (resistor). Further scrutiny also revealed the conceptualisation of resistance in terms of 'frictional' models (generating heat through frictional contact on the tunnel sides) and consumer models (both heat and current being used up at the resistor).

3.6 Practical Implications for Pedagogy: Teaching

The focus on creating a model of conceptual change that can be used to inform teaching strategy has derived from the recognition that developing qualitative understanding is essential for science making sense to the learner. If the knowledge that pupils encounter in science education is not to remain inert, in the sense of bearing no relation to causal explanation of phenomena experienced outside formal science instruction, then the mechanisms that facilitate such change must be addressed. Of relevance to this is metacognition, a process that requires the learner to examine the assumptions that underpin the cognitive explanatory frameworks that they use to explain the world. This would support them in considering the rationale basis of other explanatory models and consider the empirical evidence and intrinsic coherence on which these models are based. The value and role of metacognition in science learning is explored further in Chapter 6.

For teaching, this raises the issue of sequence in the order of acquisition of ideas that the student encounters during instruction. This is considered critical across several domains (for example, particle theory, Johnson 1998; astronomy, Vosnaidou 2001). There are implications in this regard for both teacher subject and pedagogic knowledge and the nature in which research informs practice. The attendant problems concern the fostering of the delicate symbiotic relation between research and practice in which it recognised that there are alternative conceptualisations as to what would constitute effective pedagogy in particular domains. This could involve identifying where, in specific conceptual domains, research findings from different contexts provide a range of insights that when combined together further inform sequence in both cognition and pedagogy. The work of Vosnaidou (2001), for example, whilst undoubtedly useful in mapping the prerequisite for understanding the day/night and seasonal cycle, makes no explicit reference to light and shadow formation. We have found this particularly significant in supporting learners in developing coherent, causal mechanism for explaining seasonal change in daylight length (see Chapter 6). It raises the issue of conceptual hierarchy concerning whether learners require instruction in the basics of light and shadow prior to undertaking this area of study. A critical point here is that the sequence of acquisition needs to be informed through research that attends to the learning process itself. The 'science text' subsequently created would therefore necessarily be a consequence of discourse in science learning and teaching. It raises interesting (and important) questions that need to be addressed concerning teacher subject and pedagogic knowledge.

3.7 Teacher Subject and Pedagogic Knowledge

The remit of teacher education transcends the technological enterprise of practical issues of circuit building or quantitative proficiency in the manipulating of algebraic equations to solve circuit problems. Important though this is, of itself, it does not address how best to support teachers in developing the confidence to engage effectively with children's ideas on simple circuits. If teachers are to support learning in which children will have ideas and raise questions about how electric circuits function (such as what makes a bulb light, why are two bulbs less bright, what causes the battery to go flat?), they need to address their own interpretations and qualitative understanding. The development of meaning is integral to this, and derives from such learning experience.

Teachers' concerns are often focused on the adequacy of their own explanations in responding to questions raised by pupils and their training requires that they develop knowledge of appropriate pedagogy. The Professional Standards for attaining qualified teacher status, for instance, in England (TDA 2007) require that intending teachers possess 'secure knowledge and understanding of their subjects/curriculum' as well as 'related pedagogy' and that they are able to build on 'pupils' prior knowledge, develop concepts and processes, enable learners to apply new knowledge, understanding and skills'. Subject-related pedagogical knowledge is critical in underpinning such aspirations and this necessarily includes knowledge of the relative merits of models, analogies, illustrations, practical work and other methods in securing progress. Numerous studies in this area (for example, Osborne and Freyberg 1985; Osborne et al. 1991) include accounts of pupils' own explanations and reasoning in making sense of what is happening in simple circuits. Easley (1990) reported that children' questions often focus on the mechanisms of how things work and that standard textbook explanations provided little in the way of supporting qualitative understanding in teaching and learning.

In the curriculum specification referred to above, however, there is no attempt to relate this to the subject domains. This issue is not necessarily resolved with a shift in emphasis from the focus on teachers' misconceptions towards identifying the knowledge base required for the successful teaching of simple electric circuits to primary children. Summers et al. (1998), for example, explicitly identify the subject knowledge base required as the foundation for the teaching of electricity with exemplification from a particular case study based on a particulate approach. They argue that the success of the particular analogy in question (the bicycle chain analogy) is dependent on the pupils' experience and the analogy also supports a holistic (systemic) rather than sequential interpretation of circuits. There is a degree of tension in this regarding the acknowledgement within the study as to the inherent problems that arise in supporting the idea of current conservation without consideration of a causal mechanism for energy transfer at the bulb. It is difficult to support their argument for subject-specific teaching knowledge without consideration of what could constitute a *synthesis* of subject and pedagogic knowledge in which the teacher recognises the critical learning elements within particular conceptual frameworks. Building on our previous research in this area (for example Parker and Heywood 2000), we consider the synthesis of subject and pedagogy further in Chapter 7.

An important pragmatic challenge within the remit of evidence-based practice concerns how to achieve this synthesis across the range of phenomena stipulated in the statutory curriculum for science. In addressing the problem, the research emphasis needs to shift focus from determining teachers' misconceptions towards teacher interpretation during the learning process itself. This explicitly acknowledges the teacher as principal interpreter and broadens the scope from concerns with misconceptions, based on the notion of recapitulation, towards a focus that attends to the issue of how meaning is generated in science learning and teaching discourse.

The emphases of science knowledge in science education and the recapitulation of that knowledge in which learning is perceived as a *journey towards* meaning that awaits linguistic expression focus the pedagogic task, in this case, of a search for the Holy Grail of analogies. In this model, successive analogies would be compared as to their effectiveness in explanation. This detracts from the generative power that intellectual engagement with analogy fosters, situating interpretation and meaning within the context of a *journey towards* rather than a *state of being* in the world. In the former, the 'parts' to 'whole' is problematic when an analogy breaks down, in the latter, this limitation would be perceived as integral to what *meaning* is. In this, resides the unique contribution of hermeneutic theory to the evolution of the text that is science as presented in science education discourse. It is worth considering here the relation between cognitive resonance and dissonance in the generation of meaning through analogy. While cognitive resonance is seductive, it needs to be considered with caution since the notion that concepts in electricity are uniquely apprehensible as the preceding discussion illustrates, is untenable.

The challenge that confronts both researcher and teacher concerns how teacher subject and pedagogic knowledge can be combined for effective practice (Fensham 2001). This synthesis, derived from reflection on the learning process itself (metacognition), requires the interpretation of subject knowledge for appropriate pedagogy and as such is the very exemplification of the hermeneutic task. This critical element of science learning is currently being addressed (see for example Parker and Heywood 2000; Heywood and Parker 2001) and offers considerable potential to inform the current research paradigm focused on evidence-based practice.

Chapter 4
Cognitive Conflict and the Formation of Shadows

Chapter 2 outlined some of the complexities of the conceptual change process and illustrated difficulties experienced by primary teachers in developing qualitative understanding in the domain of force and motion. We demonstrated that by adopting a metacognitive approach to learning, learners become aware of how their thinking is shaped and moulded as they interact within the social learning context. In this process, important pedagogical insight is generated to inform future practice. It affords the opportunity not only to explore the embryonic emergence of pedagogical knowledge in teacher education but also to engage with individuals' epistemological beliefs about the teaching and learning of science that have been shown to be powerful influences in shaping classroom approaches (Lunn 2002).

In this chapter, we intend to examine the conceptual change process further by focusing on the introduction of cognitive conflict to promote learning in science. Arising as a consequence of the Posner et al. (1982) model of conceptual change discussed in Chapter 2, cognitive conflict is a key instructional strategy in conceptual change studies focused on challenging students' alternative conceptions (for example: Stavy and Berkowitz 1980; Piaget 1985; Strike and Posner 1985; Thorley and Treagust 1989; Dreyfus et al. 1990; Jensen and Finley 1995; Niaz 1995; Limón 2001). We explore the efficacy of the practice by drawing on educational research in the field and through case study of pre-service teachers learning about light and shadows. We consider the conceptual domain of light to be one of the most difficult areas of physics for the non-specialist and explore some of the challenges presented by the subject matter for the learner seeking a qualitative level of understanding. The chapter concludes by considering some of the complexities of the cognitive conflict strategy and its implications for the teaching and learning of science.

D. Heywood and J. Parker, *The Pedagogy of Physical Science*, Contemporary Trends and Issues in Science Education, vol. 38, DOI 10.1007/978-1-4020-5271-2_4,

4.1 Promoting Conceptual Change Through Cognitive Conflict

4.1.1 The Role of Cognitive Conflict in Learning Science

Cognitive conflict has a long tradition as a strategy for promoting conceptual change in science learning (for example: Stavy and Berkowitz 1980; Strike and Posner 1985; Hashweh 1986; Driver 1989b; Thorley and Treagust 1989; Chi 1992; Chinn and Brewer 1993; Gorsky and Finegold 1994; Niaz 1995; Shepardson and Moje 1999; Limón 2001; Lee et al. 2003; Parker 2006). The powerful model of conceptual change, developed by Posner and his colleagues in the 1980s, portrays learning as a process by which new experience is assimilated using existing concepts, or accommodated by radical reorganisation or replacement of concepts (Hewson 1981; Posner et al. 1982a, b; Strike and Posner 1982). The cognitive conflict strategy is based on the idea that by introducing *anomalous data* or *discrepant events* into learning, learners are caused to recognise that their existing understanding cannot explain the evidence presented. In experiencing cognitive conflict, learners become dissatisfied with their understanding, and in the process of questioning the usefulness of their thinking, the way is paved for the more plausible and useful scientific concept. Conflicting experience used to trigger cognitive conflict may be in the form of direct observation, practical exploration of a phenomenon or in the presentation of incongruous ideas through written text or discussion. In Chapter 2, for example, we illustrated the centrality of weight as the dominant feature of learners' explanations for floating and sinking and illustrated how, introducing the discrepant event of observing the floating of large, heavy objects and the sinking of small, light objects, students are caused to examine their thinking in the quest for a more satisfactory personal rationale.

4.1.2 Some Limitations of the Cognitive Conflict Strategy

Researchers have used various terms together with cognitive conflict to describe situations in which learners are caused to recognise discrepancies between anomalous data and their existing reasoning including, for example, *cognitive dissonance* (Botvin and Murray 1975), *discrepancy* (Zimmerman and Blom 1983) and *paradoxes* (Movshovitz-Hadar and Hadass 1990). Although, on the surface, cognitive conflict appears to be a logical and sensible strategy to adopt in the quest to promote conceptual change, historically, it has proved to be somewhat controversial. While some studies report positive results for employing cognitive conflict in conceptual change teaching (for example, Stavy and Berkowitz 1980; Thorley and Treagust 1989; Jensen and Finley 1995; Limón and Carretero 1997; Mason 2000; Lee et al. 2003), others have found that learners may fail to recognise conflict or only partially resolve it using existing conceptions (Dreyfus et al. 1990; Dykstra 1992;

Chan et al. 1997; Trumper 1997; Chinn and Brewer 1998; Tao and Gunstone 2000). Chinn and Brewer (1998) describe seven responses to anomalous data in that people may simply ignore, reject, exclude or hold such data in abeyance until the discrepancy is resolved. Alternatively, data may be reinterpreted, or accepted with changes made to peripheral theory, while the core theory remains unchanged, or the data may be accepted and core theories changed accordingly. Simply presenting anomalous data alone is, therefore, no guarantee of falsification of existing ideas and conceptual change.

Research points to a host of factors that might influence personal response to cognitive disharmony. These include, for example, motivation, attitudes, learning strategies, cognitive reasoning abilities, the availability of alternative explanations, epistemological beliefs and socio-cultural factors (BouJaoude 1992; Chinn and Brewer 1993; Pintrich et al. 1993; Chan et al. 1997; Hofer and Pintrich 1997; Mason 2000; Park et al. 2001; Kang et al. 2005; Palmer 2005). Indeed, Strike and Posner's later extensive revision to their original theory of conceptual change considered the theory to be over rational in that it understated the range of factors that might function as part of the conceptual ecology influencing conceptual change (Strike and Posner 1992).

In considering the efficacy of cognitive conflict as a strategy for promoting conceptual change, the practice cannot be extricated from the socio-cultural context constituted by individuals interacting within real learning environments. Limón (2001), for instance, emphasises the importance of securing *meaningful* cognitive conflict such that learners should be motivated, have their prior knowledge activated and possess adequate reasoning abilities and appropriate epistemological beliefs. She suggests that clearly highlighting the difference between students' beliefs and contradictory experience can help students to better reflect upon their existing conceptions as they attempt to explain and rationalise the conflict. Kang et al. (2005), in a study involving seventh grade Korean students, found that their abilities to select necessary information and perceive it as separate from their existing conceptions (field dependence/independence) is a significant factor in the degree of cognitive conflict induced. Other cognitive aspects may also be influential. Zohar and Aharon-Kravetsky (2005) reported that students with higher academic achievement benefited more from a cognitive conflict approach, whereas lower achievers benefited more from direct teaching methods designed to diminish ambiguity in learning situations.

Despite its controversial nature, cognitive conflict is still considered a valuable strategy in inducing conceptual change and we would contend that it addresses an important function in education; it highlights the unexpected nature of the way in which science describes how the world works and promotes critical thinking as an important part of science learning. It also necessitates engagement with learners' ideas and explanations as a key feature of effective science education. In this chapter we present empirical evidence of pre-service teachers reacting to cognitive conflict contexts and consider further what might constitute meaningful conflict that can be used positively in teaching and learning science.

4.2 The Challenge Presented by the Conceptual Domain of Light

Light constitutes one of the most demanding areas of learning in physical science, probably because its production, propagation and interactions with materials and objects in the environment are imperceptible to the observer. Like other physical phenomena, such as electrical circuits or magnetic attraction and repulsion, the observer sees only the manifestation of the instantaneous process of light production, light travelling and light interacting, and relies ultimately on visualisation as a means of developing causal explanations for its behaviour. As Galili (1996) observes, current scientific thinking about light is the result of more than 2,000 years of scientific research culminating in explanations that are sophisticated, complex and non-plausible. Its physical parameters (such as speed and image formation) are beyond human perception, yet are generally represented by science and for learners of science, as stationary and continuous (Galili and Hazon 2000). Thus, if students are to develop *understanding* of optical phenomena, they will need to synthesise knowledge concerning light propagation, illumination pattern and the observer's visual pattern (Langley et al. 1997). In the examples of learning that follow, we explore how pre-service teachers developed qualitative explanation of shadow formation; a process requiring a complex synthesis of a range of knowledge including the following:

- light as an entity produced by a light source
- light as an entity propagated in space
- reflection and absorption of light as it interacts with an object
- shadow formation as an area of darkness that varies in intensity according to how much light is reflected from the object into the eye
- the role of the eye as a receptor

At the primary level of education, the study of light tends to be rooted in experiential learning that focuses on common manifestations of light such as shadow formation. In the national curriculum of New Zealand, for instance, shadow formation is identified as a physical phenomenon suitable for the learning of primary-aged children in physical science (MoE, NZ 2007). In English primary schools, the science national curriculum focuses on the identification of light sources, light travelling, light interaction with materials, light reflection and seeing being the result of light entering the eye (DfEE/QCA 1999). Although teaching tends to concentrate on experiential learning rather than explanation of what light *is*, underpinning common practical activities such as exploring shadow formation and reflection lies the notion of light as an *entity* travelling in space and interacting with materials it meets to produce certain observable effects. In accounting for these events, the primary teacher is faced with the challenge of how best to support young learners in rationalising their observations. The teachers' knowledge and understanding of the subject are important because they underpin decisions about the inclusion of knowledge bases in the curriculum that might otherwise seem arbitrary to the non-specialist teacher. Teaching that lacks such a coherent rationale will invariably fail to recognise opportunities and pathways for pupils' conceptual progress. Developing a qualitative understanding of a phenomenon such as shadow formation therefore entails engaging teachers in examining fundamental notions of how light is produced, propagated, received by the eye and interpreted by the brain, as well as how it interacts with materials encountered as it travels.

4.3 Exploring the Impact of Cognitive Conflict in Learning About Shadows

4.3.1 Background to the Exemplification Study

The following discussion draws on an empirical study by Parker (2006) that examined the effect of cognitive conflict as an instructional strategy in primary pre-service teachers' learning about light and shadows. In particular, it focused on how learners perceived and responded to a range of increasingly difficult scenarios designed to challenge their thinking about shadow formation, and how they tried to resolve the conflict through experimental activity, reading and discussion. In order to support critical reflection, students documented their personal cognitive journeys in a learning journal. The role of the tutor was to help learners to:

- become aware of their spontaneous explanations of shadow formation;
- recognise when and how personal conceptions are influenced during the learning process;
- become aware of when and why personal conceptions no longer explain outcomes to satisfactorily;
- explore shadows by challenging thinking with 'what happens if…?' type questions;
- focus observation on the location, size, shape and intensity of the shadow.

The participants comprised 13 primary undergraduate pre-service teachers studying science as a main subject option. Only one of the groups had previously studied physics post 16. In the final year of their initial professional training, the group was comfortable with a learning approach based on the sharing of ideas and reflecting on learning experience. This is a critical factor in influencing personal reaction to conflict situations that must not be underestimated in the quest for meaningful engagement. Learners' responses were documented in regard to a range of scenarios involving an increasing degree of cognitive conflict in relation to shadow formation. A metacognitive approach (detailed in Chapter 1) was adopted to support learners in identifying (or formulating) their existing explanations, recognising when personal reasoning conflicts with a discrepant event, and questioning, investigating and critically reviewing evidence in the quest to develop qualitative, causal explanations for shadow formation. Qualitative data comprising in this study were drawn from interviews, discourses, reflective journals and summative assignment writings (see Parker 2006 for detail). The data presented here are used as exemplification, presented both numerically and qualitatively following a 'critical case' sampling process (McMillan and Schumacher 2001) in which subjects are selected because they are representative critical examples of student learning.

Students were presented with three consecutive learning contexts involving increasing degrees of cognitive disturbance. Prior to practical investigation of the scenarios, they made predictions of outcomes in terms of shadow location and appearance (for example, size, depth of colour), and subsequently noted points of interest and questions arising as journal entries to assist them in identifying their own learning.

4.3.2 The Cognitive Conflict Scenarios

4.3.2.1 Scenario 1 – Two Light Sources, One Object (Fig. 4.1a)

This activity provided a starting point designed to orientate thinking about shadows and stimulate discussion of how shadows are formed. It was anticipated that, as most people's past experience of learning about shadows tends to be based on situations involving a single light source, introducing a second light source would increase the potential for conflict in reasoning. Investigation focused on shadow formation with both lights and individual lights switched on.

4.3.2.2 Scenario 2 – Multiple Light Sources (Fig. 4.1b)

This constituted an extension of scenario 1 in that it employed multiple light sources (five bulbs arranged 4 cm apart) and permitted investigation of a range of light source possibilities. As students are unlikely to have experience of focused observation of the shadows produced by multiple sources in detail, it was anticipated that this would represent a greater cognitive disturbance than scenario 1.

4.3.2.3 Scenario 3 – Using a Cross-Shaped Light Source (Fig. 4.1c)

Developed from the work of Rice and Feher (1987), this scenario involved a crossed-shaped light source. When a small bead is suspended in front of the light source, the cross-shaped shadow produced is dramatically surprising and highly counterintuitive to most learners. It is likely to cause significant cognitive disturbance as a consequence of the intuitive association between shadow formation and object that fails to consider the light source. Two aspects are explored here:

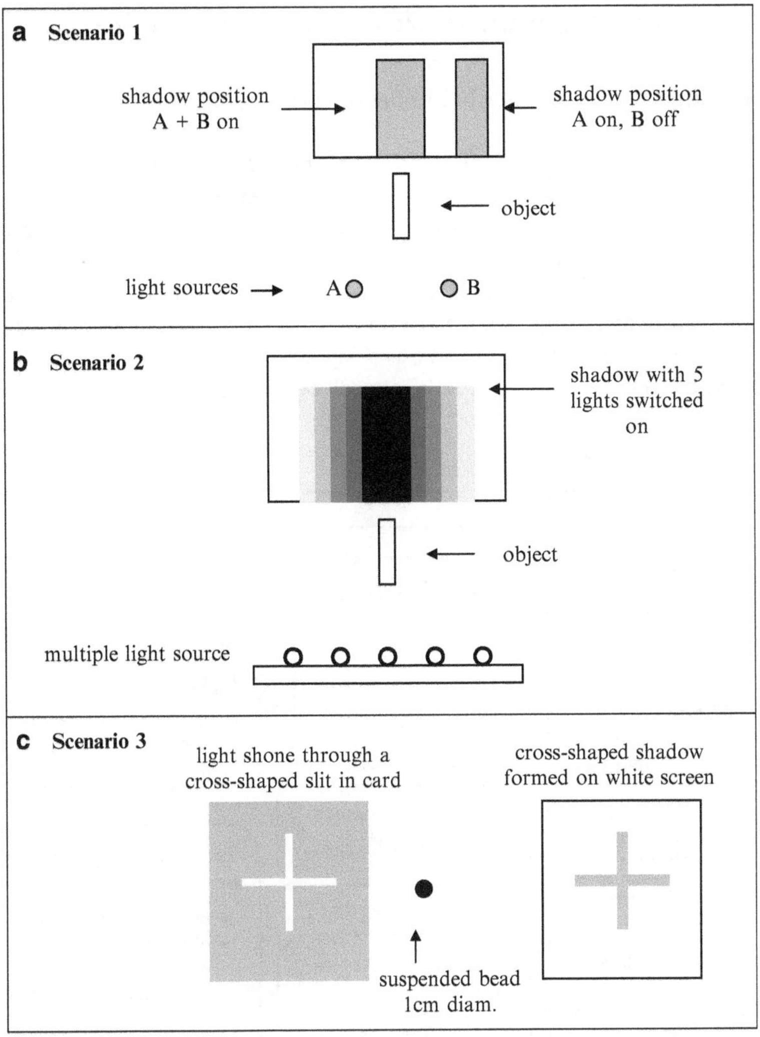

Fig. 4.1 Using a cross-shaped light source to illuminate an object

1. The pattern of illumination produced on the screen by the cross-shaped light source
2. The shadow formed on the screen when a small bead is suspended in front of the light source

In each context students predicted the location, shape, size and intensity of the shadow prior to investigation and outcomes were examined critically in respect of their initial explanations for shadow formation. The role of the tutor was to facilitate observation of the scenarios and to support students in articulating their evolving reasoning.

4.3.3 Learner Responses to the Cognitive Conflict Scenarios

4.3.3.1 Initial Conceptions About Shadow Formation

Scenario 1 helped to orientate thinking within the context, and provided a vehicle for the articulation of ideas about how a shadow is formed. Figure 4.2 summarises the body of knowledge expressed and shows that learners tend to focus on how light travels, what happens when it meets an object and how this leads to the formation of the shadow. Light is conceptualised variously as travelling in straight lines, as 'rays' or 'beams' and being unable to travel around objects. The shadow is generally viewed as light being 'blocked' from travelling through the object, thereby causing an area of 'darkness' where light fails to reach behind the object.

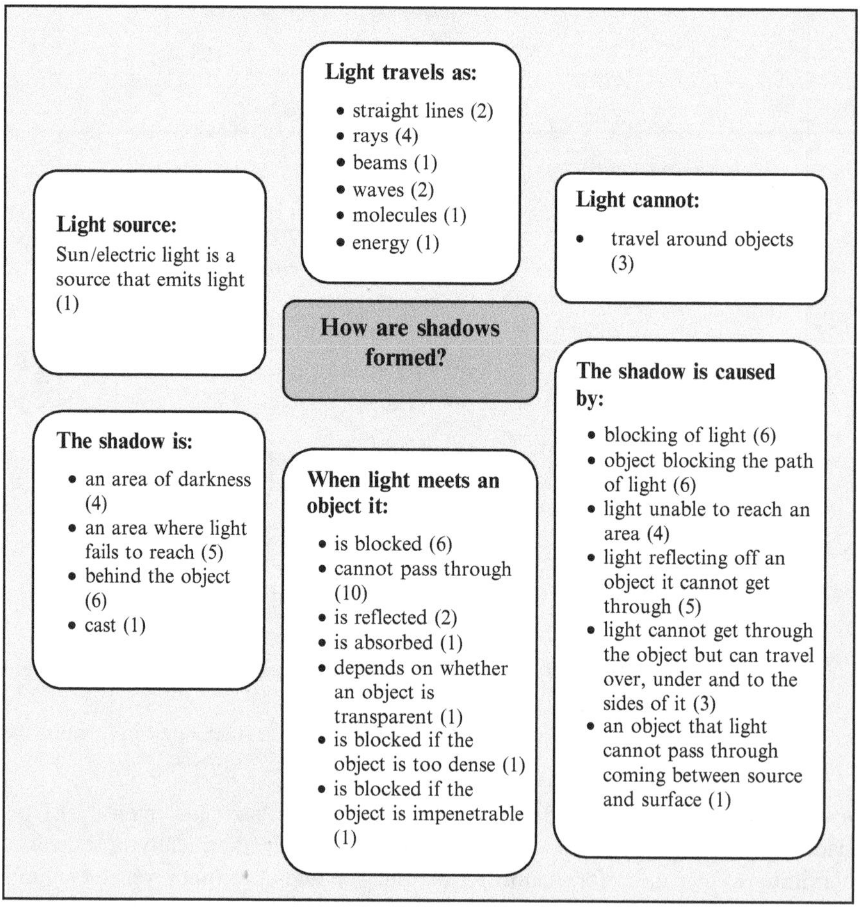

Fig. 4.2 A summary of students' intuitive knowledge prior to teaching (N = 13, numbers of responses in brackets)

Science education research has amassed a considerable bank of knowledge pertaining to learners' scientific conceptions, including ideas about light and various optical phenomena (see Duit 2008 for a bibliography). In a study of high school (tenth grade) and college (teacher-training) student conceptions of light, vision and related topics, Galili and Hazon (2000) suggest a hierarchical structure described in terms of facets and schemes. In regard to shadow formation, they propose a *Shadow Image Scheme* (SIS) comprising ten facets of knowledge, in which shadows are perceived holistically, in much the same way that optical images are frequently viewed. This scheme is robust, changing little after instruction. Learners frequently view shadows as *entities* that can be manipulated, added and subtracted. The shadow is viewed as representing the object's shape and the role of light is to make the shadow 'visible'. Shadows are reified like images in mirrors, and such a model cannot explain the occurrence of partial shadows. Several elements of the pre-service teachers' reasoning resonate strongly with the facets of knowledge constituting Galili and Hazon's *Shadow Image Scheme.* Students attributed the mechanism of shadow formation mainly to light being blocked by an object through which it is unable to pass. Some recognised light being reflected or absorbed by the object, and generally the shadow was seen as the area where light did not reach (an area of darkness). Six students mentioned that the shadow would be located 'behind' the object and four referred to it as being 'cast'.

Figure 4.2 shows that scant attention is given to the light source in reasoning (a facet of the SIS is that the shape of the shadow is not dependent on the light source). Thus, although students expressed the notion of shadow formation caused by the blocking of light, it seems that a high degree of their reasoning is predicated on a holistic view of shadows. Predictions and explanations frequently portrayed shadows as entities that can be manipulated. They referred, for example, to a 'shadow behind a shadow' or 'shadows moving' when one light is turned off. Also, none of the students predicted the occurrence of partial shadows (another facet of the SIS), and this was to become a focus of their thinking during investigations, as they were unable to explain the phenomenon satisfactorily within existing frameworks.

In interview, all but one individual subscribed to a model of light production and propagation whereby light rays (also described as 'beams' or 'waves') are emitted in straight lines in all directions from a central point in the light source (see Fig. 4.3a). Although one person expressed the scientific view of light being emitted in all directions from each 'point' on the light source (Fig. 4.3b), the student was unable to apply the model correctly in predicting the nature of the shadows formed in scenarios 2 and 3.

4.3.4 Categories of Responses to the Cognitive Conflict Scenarios (1–3)

Four distinct categories of response could be discerned:

1. The shadow(s) formed is simply described in terms of its location, size, shape and/or intensity. Responses generally place emphasis on an unanticipated

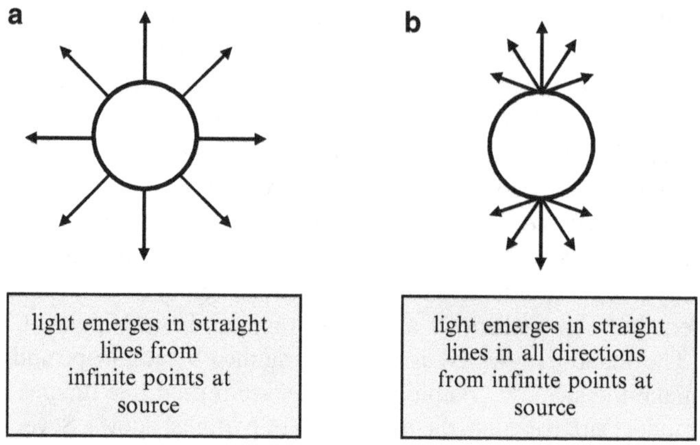

Fig. 4.3 Two different perspectives of light production at source

element of the shadow produced, such as partial shadows frequently described as the 'hazy' or 'blurred' parts of the shadow (category 1).
2. Generalised 'why' questions arise in response to unanticipated element(s) of the shadow. These indicate what the learner has found surprising in regard to the shadow observed. For example, a student may simply ask why some parts of the shadow are 'hazier' than others (category 2).
3. Hypotheses are formulated about the relative positions of the light source(s) and object as a means of explaining outcomes. This reflects Galili and Hazon's the *Shadow* Associative Scheme in which students associate key aspects of the phenomenon (such as size, strength, location and inclination of light source), in an attempt to develop an explanatory mechanism to make sense of reality by establishing cause and effect between the subject (shadow) and environmental parameters (category 3).
4. Hypotheses based explicitly on light production, propagation and the mechanism of shadow formation are formulated in order to provide a causal explanation for observations. Such a response is indicative of a much deeper level of thinking and may contain evidence of learners actively questioning their understanding of how light travels outwards from the source and how this results in the shadow formed (category 4).

The categories of response provide a crude measure of the degree of cognitive disturbance produced by the scenarios as they represent a movement from beginning to notice differences between expectation and observation (typified by category 1/2 response) towards engaging in associating environmental parameters (category 3), or beginning to question underlying concepts (category 4).

4.3.4.1 Responses to Scenario 1

In scenario 1 (Fig. 4.1a), only 2 of the 13 participants predicted correctly that two shadows would be produced when both bulbs were switched on. Consequently, for most of the group, this represented a discrepant event. However, responses to the potential cognitive conflict were largely of category 1/2 type and focused on the number, shape and size of shadows formed as well as the occurrence of partial shadows as the following category 2 responses illustrate:

> Why are there two shadows with a gap and not just one?
>
> Why is it fuzzy at the top?

Three students speculated that the shadows were formed as a consequence of the relative positions of light source and object (category 3 response):

> If both light sources were angled would we see only one shadow?

The absence of category 4 responses indicates that although some elements of this scenario such as the occurrence of two shadows and partial shadows had prompted mild cognitive disturbance, this was insufficient in stimulating thinking. Students either rationalised the experience within existing schema, possibly making peripheral changes to core schema to assimilate new observations, or simply ignored the discrepant evidence. Such responses accord with Chinn and Brewer's (1998) categories of response to cognitive conflict.

When light source A (Fig. 4.1a) was switched off, all participants predicted correctly the formation of a single shadow, reinforcing past experiences of learning where only one light source is typically used in considering shadow formation and is, therefore, unsurprising in that sense. However, the position of the shadow, its intensity, and the occurrence of partial shadow comprised key elements of their category 1/2 responses:

> Why did the shadow opposite the light source disappear?
>
> Why is there a hazy bit around the top of the shadow?

Three students engaged with category 3/4 responses:

> The angle of the light source determines shadow formation.
>
> It is less bright when the light is turned off and therefore the shadow is darker.
>
> When the light is turned off the shadow moved and blended with the other to give one of greater intensity. It looks like a shadow behind a shadow.

Further insight into developing causal explanation is revealed through discussions between student (ST4) and the tutor:

> Tutor: Can you explain what your diagram means in terms of how the shadow is formed?
> ST 4: The same shadow is formed if A is on, B off or if A is off, B on. When A is off and B is on the shadow will be formed directly behind B as the card stops the light from travelling, if it were the other way round B off and A on, then it would be formed at the other side of the screen.
> Tutor: Why have you drawn the two lines going from B to the shadow like that?
> ST 4: They show light travelling from the bulb and hitting the object but they can't pass through it just over, under, to the sides of it.
> Tutor: Can you show me how you think of light travelling out from the light source?

ST 4: [draws diagram as in Fig. 4.3a]. The rays come out in all directions.
Tutor: Do you mean there's one ray from each point the light source?
ST 4: Yes, but there are millions of them, they go in all directions and it's only the ones that get blocked that make the shadow

This student is equating each 'point' on the light source with the emission of a single light ray, as is the following student:

Tutor: Can you tell me what your diagram means?
ST 7: The same shadow is formed if A is on, B off or if A is off, B on. When A and B are on it might be bigger (Fig. 4.6b). The shadow is formed when something opaque is blocking the light.
Tutor: When A is on and B is off why would the shadow be in the middle of the screen?
ST 7: Because the shadow is always behind the object because the light goes in straight lines.
Tutor: Is that how it travels from the bulb?
ST 7: Yes it travels in straight lines spreading everywhere.
Tutor: Like in this diagram (Fig. 4.3a), just one ray from each point?

Thus, the same conceptions of light production that underpin reasoning about shadow formation resulted in two different predictions. Students did not spontaneously engage with notions of light production from the source unless interviewed specifically on the subject. Moreover, light sources feature infrequently in intuitive thinking about shadow shapes (Fig. 4.2), instead it is the underlying conception of light production and propagation that informs prediction.

4.3.4.2 Responses to Scenario 2

As anticipated the multiple light source scenario (Fig. 4.1b) provided greater depth of cognitive conflict and consequently seven individuals were unable to even formulate predictions about the nature of the shadow that would be formed if all light sources were turned on simultaneously. Following investigation, seven students produced category1/2 responses such as:

The darkest shadow is in the centre surrounded by many shades, four on either side.

Why was there a different shadow in the centre when the light source was exactly the same with no central main source?

After systematically switching off individual lights, two students proposed a quantitative relationship between source and shadow (category 3 response):

If there are 6 bulbs will there be 6 shadows, 12 bulbs and 12 shadows and so on?

Four individuals began to engage with causal explanations (category 4 response):

The lights must be creating their own shadows. These will all be the same intensity but because the lights are on at the same time some of the shadows overlap causing different light intensities, the places where the shadows overlap make a darker shadow. If more shadows overlap the darker the shadow will become.

> Why does the shadow appear to create its own shadow of the shadow? The shadows appear to create its own shadow, but because there are other bulbs shining on the object from different angles, some shadows are hazier.

The cognitive disturbance had been more successful in focusing some students on the light source with regard to shadow formation, and they recognised this explicitly as being counterintuitive:

> The shadow does not seem to be solely related to the object and this goes against what I have always thought. Shadows in everyday circumstances don't look like this despite several light sources.

Although the activity succeeded in moving students towards beginning to incorporate the light source within their conception of shadow formation, they still reasoned about shadows within existing paradigms of light production and propagation. They explained the outcome by splitting the shadow into parts, each created by a single light source, and then rebuilding the whole from the constituents. It had not engaged them in thinking further about light production and propagation and existing schema remained intact.

4.3.4.3 Responses to Scenario 3

For all of the learners in this study scenario 3 (Fig. 4.1c) constituted a dramatically disturbing phenomenon and the entire group predicted that the cross-shaped light source would produce a cross-shaped area of illumination on the screen. When this failed to happen, and the screen became fully illuminated, they were highly surprised. Furthermore, on observing the cross-shaped shadow produced by the suspended bead, in stark contrast to their expectation of a round or oval shadow, the reaction was frequently one of disbelief. Immediate responses were dramatic:

> Wow – why?

> This was a complete shock – why is it in textbooks we see shadows, as the place where light is blocked yet why is the shadow spread into a cross?

Although all immediate responses were of category 1/2 type, these were fleeting and soon replaced by category 4 responses as all students began to question their own causal mechanisms of shadow formation in regard to the role of the light source and light production and propagation:

> Surely if light travels in a straight line it doesn't only come out at 90° like would be expected from the results?

The cross-shaped shadow proved so counterintuitive that it resulted in a number of questions significantly different in nature from those of the preceding activities indicating that long-held, fundamental beliefs about light production and propagation were being questioned. The questions are indicative of a deep level of thinking about science concepts and signal the opportunity for conceptual development.

In Chapter 3 we illustrate how the failure of analogies to provide satisfactory explanation for all aspects of the behaviour of the simple electrical circuit affords the opportunity to raise questions and extend learning. There are similarities with the cognitive conflict situation described here and both present opportunity for learning to occur. Also notable at this point is the strong affective nature of the student response and that varied from stimulation of interest and the creation of a strong drive to find out more in some individuals, to an expression of concern, lack of confidence and anxiety in others. Managing this tension is a critical part of learning on behalf of both teacher and student and it will be discussed at greater depth later in this chapter as we consider issues related to resolving the conflict.

4.3.5 Triggering Meaningful Cognitive Conflict

Chi (1992) emphasised the need to understand which triggering factors could lead students to realise the anomalies to be resolved. Additionally, Park and Pak (1997) argue that to stimulate cognitive conflict students must be able to discriminate evidence from their own ideas and to recognise whether or not this evidence is available to support or disprove their own idea. Although scenarios 1 and 2 in this study had failed to trigger an explicit conflict, they raised questions for some individuals about the relationship between light source and shadow production, whilst for others they reinforced existing ideas about how shadows occur. Either way, they allowed students to activate (or, indeed generate) their existing understanding prior to the introduction of the counterintuitive third scenario that triggered cognitive conflict for all the students in the group. Although getting students to recognise the conflict was problematic in scenarios 1 and 2, this was eclipsed in scenario 3 by the dramatic appearance of the cross-shaped shadow. However, this constitutes no guarantee that the students will engage *meaningfully* with the conflict such that learning takes place. It might simply be categorised as a peculiar exception to what is normally perceived and simply associated with the unusual light source used.

4.4 Resolving the Conflict

4.4.1 The Need to Generate Causal Explanation

In seeking to promote conceptual change, it is clearly insufficient merely to disturb thinking through the introduction of anomalous data. Driver (1989a, b) claimed that providing the opportunity to refute prior misconceptions is insufficient because such misconceptions will not be rejected until there is something reliable to replace them with and that learners must have opportunities to generate alternative interpretations. Park et al. (2001) support this contention in a study of using cognitive conflict with middle school and college students to promote understanding of electrostatic induction. They found that contradiction alone, without the opportunity to generate any new alternative explanatory hypothesis, could not falsify the core of students' conceptions. Once initial curiosity has been aroused through the introduction of discrepant data, learners' motivation to resolve the conflict plays a key role in learning. Research shows that the extent to which a single discrepant event will maintain student motivation is debatable (Hidi and Harackiewicz 2000). The resolution of disharmony through the emergence of meaningful understanding is critical if students are not to be left with inappropriate interpretations, in a state of confusion or simply disengaged from learning. So strong is the need to resolve conflict that students may do so by modifying their observations rather than their understanding (Gunstone 1991). Consequently any approach designed to support learners in resolving conflict needs to help individuals perceive clearly the key elements that cause such conflict in the quest to maintain motivation to find resolution. The aim is to facilitate construction to the scientific view in such a way that learners build meaningful, causal explanations for events and this needs to be differentiated from outcomes where learners may resolve conflict to their satisfaction but do so by developing inappropriate learning outcomes (Fensham et al. 1994; Baddock and Bucat 2007).

4.4.2 Resolving the Cognitive Conflict Caused by the Cross-Shaped Shadow

The Parker (2006) study sought to support students in resolving the conceptual disharmony created by the cross-shaped shadow by employing a structured investigation involving only the vertical component of the light source in order to reduce the complexity of the shadow. In simplifying the light source, the intention was to develop understanding of the whole by a process of reconstruction. Figure 4.4a shows that as the bead is moved away from the screen and nearer to the light source, the emergent shadow changes from what might be approximately bead-shaped (at the 2 cm position) to a vertical form (at the 12 cm position). At positions 2, 4, and 12 cm from the screen students made predictions of shadow location, size, shape and intensity and explored the outcome.

All participants recognised this as the vertical component of the cross-shaped shadow, and most students predicted the shadow would get longer as the object moved further from the screen and towards the light source. However, this prediction was largely based on the experience of having seen the cross-shaped shadow rather than personal reasoning. The longer the shadow became and the less it resembled the bead in shape, the more students struggled to rationalise it:

> I was wondering if the shadow was like that because the ball is shiny, making it a bit blurry with lighter and darker parts but if you put it closer to the screen it will be clearer – so it must be the distance from the screen being small. *(2 cm position)*

> Is the shadow related to the source? The object? Both? Why is this different from the cross light shape? Does the type of bulb matter? *(4 cm position)*

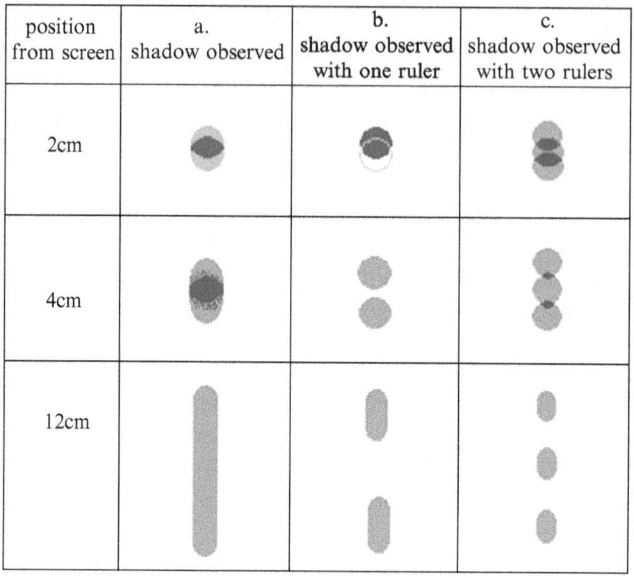

Fig. 4.4 Shadows produced by placing the bead at 2, 4, and 12 cm from the screen

> You can't tell from the shadow that it is a bead. *(12 cm position)*

The ability to extract key features of the context from all the sensory information received is an important step in beginning to resolve conflict and other studies have shown that there is need to explicitly support learners in this process (Johnstone 1997; Baddock and Bucat 2007). Comments such as those above indicate that as students contemplate shadows produced with the vertical light source, they are beginning to recognise some key elements in regard to light source, distance from the screen and shadow shape.

Although observing the change in shape of the shadow as the bead moved nearer to the light source served the purpose of focusing learners on the relationship between the two, it did not provide explanation. Consequently, in order to deconstruct the shadow further, one and then two rulers were slowly moved down the light source and observations of the effect on the shadow were made (Fig. 4.4b and c). This experience proved to be critical in beginning the process of resolving the conflict and it served to focus attention on how light travels from the source. At first, there was surprise that there was no distinct shadow of the ruler(s) on the screen, but students moved on to consider that the effect of placing the rulers was to partition the light and create two (in the case of one ruler) and three (in the case of two rulers) light sources, each producing a shadow of the bead. The activity produced a number of category1/2 responses and students identified a key element in regard to the direction of light travelling from the source:

> When the top part of the shadow is blocked the bottom part of the shadow disappears, the light must be travelling at an angle.

They began to postulate a relationship between the number of rulers applied and shadows produced:

> Why do 3 circles appear, would there be 4 with 3 rulers?

Some students drew their own representations of how light was travelling from the source to form the shadow and an example is presented in Fig. 4.5. Students typically speculated about splitting up the light source into more parts and how the shadow might be reconstructed:

> If you split the light source up further there would be many overlapping images. Are we splitting the light source up into multiple sources, each produces their own shadow?

A range of learning ensued varying from assimilation of the light source into students' conceptual frameworks for shadow formation, to radical reorganisation of thinking about how light travels from a source for nine of the group. Learning encompassed:

- considering the distance between the light source, object and screen and speculating about the resultant size, shape and intensity of the shadow produced;
- reasoning that the vertical shadow can be considered as multiple, superimposed shadows when the light source is partitioned;
- being able to construct the horizontal component of the cross-shaped shadow through extrapolation from the vertical shadow;

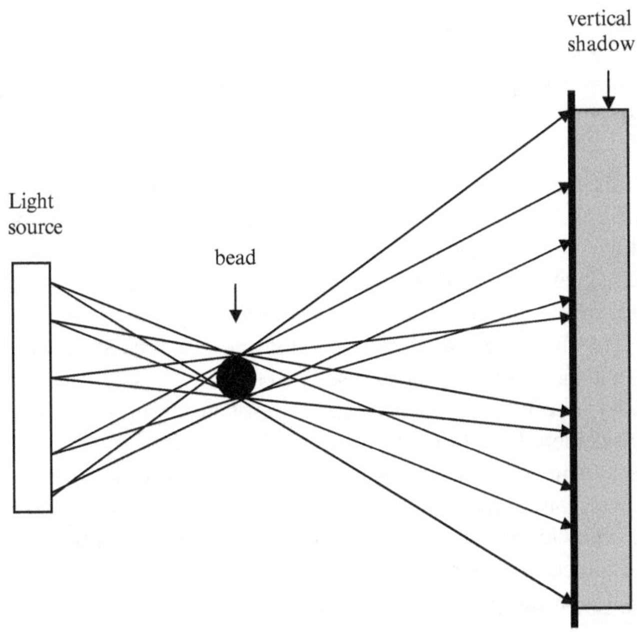

Fig. 4.5 Student representation of vertical shadow formation showing shadow formation from a vertical light source showing five points of emission

- considering how light is being emitted from the source and how it is propagated in space.

Such learning is evidenced in written responses such as:

> I now have a much better understanding of the cross shadow because I can imagine all the infinite bits of light coming from the cross light and hitting the bead in all directions rather than just seeing it as one source. I now see it as having different parts, which make up the shadow and they all add together to make the image.

Of course, because of the intensity of the cognitive disturbance produced by this phenomenon, immediate responses still contained a large degree of uncertainty. Students continued to contemplate their findings through research, including texts on both children's ideas about light production and propagation (see for example, Guesne 1985) and the explanation of the cross-shaped shadow by Feher and Rice:

> Each point of the light source emits light in all directions. Each ray travels in a straight line from the source to the screen unless an opaque object blocks it. In the limiting case of a tiny bead, only one ray is blocked for each point in the cross light. Thus there is a point-by-point correspondence between the shadow and the source. (Feher and Rice 1988: 638)

Discussion of such texts served to maintain motivation to resolve the conflict. Written reflections in the form of an assignment analysing personal learning experience revealed considerable insight into student interaction with representations of light:

> The main problem that I had with the drawing of ray diagrams is the seemingly simple diagrams used to represent a very complicated process. As I now understood it light, is comprised of infinite rays of light being emitted in straight lines in infinite directions from infinite points on the light source. How had someone managed to simplify all these angles and rays into one stationary diagram with only a couple of rays bouncing around? After examining some of these diagrams in textbooks, it became clear that only the extreme light rays were recorded on the diagram. By this I mean only the very most important light rays, those that are responsible for creating the visible boundaries between light and dark.

In making this recognition, the student in question proceeded to generate the diagram in Fig. 4.5, constructing a revised explanatory hypothesis for the cross-shaped shadow.

Park (2006), in exploring college students learning about electromagnetic induction during an introductory physics course, proposes four stages in generating explanatory hypotheses. The first stage involves observing the target experiment and making observations and the second is characterised by beginning to seek causal explanations by asking 'why' questions. These two stages resonate strongly with the category 1 and 2 responses identified. In the third stage of Park's framework, students search for hypotheses (using either background knowledge or experimental evidence), and this accords with the category 3 response wherein students typically combine their background knowledge of shadow formation with elements of the current experimental situation to explain the nature of the shadow formed. Park's fourth stage culminates in the generation of hypotheses. He found that not all students reached this stage and where hypotheses were generated they were of two types. *Auxiliary hypotheses* were those in which prior ideas were preserved and the student attempted to attribute reasoning to the experimental context and *theoretical hypotheses* involving theoretical causal relationships. Students' background knowledge appears to be a powerful element in generating theoretical causal explanation as the example of the student above searching her own understanding of the meaning of visual representation of light rays illustrates.

4.5 The Emergence of Pedagogical Insight

Figure 4.6 summarises the pedagogical knowledge that emerged as a result of the group developing metacognitive awareness of their learning within the context of shadow formation. Of course, not all students demonstrated all aspects of pedagogy presented; the summary represents knowledge accumulated from both individual and group perspectives. This body of metacognitive awareness is useful in fuelling further development and providing a framework for future discussion. Loughran (2002) contends that it is not helpful to regard pedagogic content knowledge (i.e. knowledge relating to the teaching and learning of specific subject matter) as residing solely in individuals, and that complimentary aspects are revealed through working with groups. The development of both general pedagogic knowledge and pedagogic content knowledge will be discussed at length in Chapter 7. For the purposes of this discussion, emergent pedagogic knowledge is analysed in regard to the learning process in general, to learning about light and to the implications for classroom practice.

4.5.1 The Learning Process

The experience of being unable to provide meaningful explanation for the cross-shaped shadow served to highlight for students the centrality of learners' conceptions and the need to address them in teaching and learning science (Fig. 4.6a.i). Through experiencing cognitive conflict they questioned their own explanations of light production and propagation and acknowledged the affective nature of such a process (Fig. 4.6a.v). When learning fails to secure the learner in developing meaningful explanation it is likely to be superficial and temporary (Gilbert et al. 1988; Woodruff and Meyer 1997). Park (1997: 332), in considering the conceptual difficulty of developing explanations for the behaviour of light, comments, 'science cannot always provide learners with commonsense explanations'. We contend that explicit acknowledgement of this in teaching affords the opportunity to engage the learner in consideration of the nature of scientific explanation; an important aspect of developing epistemological beliefs.

Practical investigation was widely recognised as crucial in providing evidence that allows the scrutinising of thinking about light (Fig. 4.6a.vii). Several students alluded to the 'wow' factor provided by the cross-shaped shadow as being significant in engaging thinking initially and providing the motivation to discover answers:

a. The learning process in general	b. Learning about light and shadows	c. Implications for future practice
i. Learners' preconceptions play an essential role in developing understanding in science. ii. Cognitive conflict provides the opportunity to identify and engage with learners' conceptions. iii. Coherent, causal mechanisms that explain observation are necessary for developing understanding in science. iv. Cognitive conflict may produce a differential response in learners. v. Cognitive conflict entails an affective dimension that impacts upon learner motivation and willingness to question existing concepts. vi. Changing ideas takes time. vii. Practical investigation is a valuable means of testing ideas. viii. Models and diagrams require careful scrutiny and interpretation. ix. The tutor plays an important role in questioning, focusing observation and discussion. x. Reading plays an important role in encouraging thinking about ideas. xi. Learning contexts need to be non-threatening and supportive in order to encourage discussion, questioning and sharing of ideas. xii. Learners need to be personally involved and experience a sense of achievement.	i. The abstract and counterintuitive nature of concepts associated with light. ii. The instantaneous manifestation of light propagation patterns obscures cause and effect. iii. The problem of defining 'what light is'. iv. How light is produced and travels constitutes a major source of misconception. v. The language used to describe light and shadows can contribute towards misconception. vi. There is a need to consider the role of the observer in optical phenomena. vii. Representations of light in teaching and learning can contribute towards the formation of misconceptions.	i. The need to develop a more in-depth knowledge of: • subject knowledge; • curriculum requirements; • children's conceptions; • techniques for promoting conceptual change; • scaffolding learning in specific domains. ii. The need to consider in planning: • how to motivate and engage children's thinking; • appropriate explanations; • key teaching points; • learning styles; • thinking time; • creating a sociocultural environment conducive to questioning and discussion.

Fig. 4.6 A summary of pedagogical content knowledge generated during learning about shadows (N = 13)

> I enjoyed the various investigations as they were obviously designed to challenge my thinking. This gave them a certain 'wow' factor that engaged my thinking immediately. I felt compelled to want to discover answers and the fact that my predictions were wrong when I tested them drew me in further.

When teachers present learners with a situation of potentially dramatic cognitive conflict it is important to realise that the conflict may simply be too difficult to contemplate and may be classed by the learner as a 'peculiar case' resulting in a barrier being created to further development of thinking. Addressing this in practice depends on the pedagogic skill and subject knowledge and understanding of the teacher in recognising the response and providing appropriate support in resolving the tension. In the case of the cross-shaped shadow, this was achieved by providing the opportunity for students to engage with the simpler vertical light source that could be deconstructed further using rulers. Conceptual 'shift' did not seem to occur instantaneously; rather, it was preceded by a period of orientation and followed by a period in which concepts were explored further through personal reflection and reading (Fig. 4.6a.x). There are differences of opinion as to whether conceptual change is evolutionary or revolutionary in nature (Wiser and Amin 2001), but since the revisionist theory of Strike and Posner (1992), there is a general consensus in science education that conceptual change is a slow revision, rather than a paradigmatic shift (for example, see Vosnaidou et al. 2001). In this study, the pre-service teachers engaged in reading encompassing not only physics subject matter, including the aforementioned Feher and Rice (1988) explanation, but also consideration of children's ideas about light (Guesne 1985). This afforded the opportunity to identify similarities and differences between personal ideas and those of other learners as well as scientific explanation.

The socio-cultural environment was widely recognised as playing a critical part in learning, and interactions between learners and tutors were paramount in stimulating discussion and focusing observation and investigation (Fig. 4.6a.ix,xi). Personal 'involvement' in the learning process provided ownership and was seen as leading towards a 'better' understanding and the sense of achievement gained was clearly a significant affective factor (Fig. 4.6a.xii):

> The actual dawning of understanding was really quite phenomenal and the resulting emotions of fulfilment were extremely satisfying and will remain with me. Having struggled through the whole process to make my conclusions I now know that the answers I found are explanations I will never forget.

Motivation is generally seen as a key feature in promoting effective learning and achievement and has formed a focus for science education research in recent decades (for example, Kaplan and Maehr 1999; Pintrich 2000; Ryan and Deci 2000; Deci et al. 2001; Palmer 2005). Palmer contends that although the 'classical' model of conceptual change describes the *conditions* necessary for change to occur, it does not contain a component for the motivational forces necessary to drive the process. In considering intrinsic motivation (i.e. doing something that is inherently interesting or enjoyable) Palmer draws attention to Lepper and Hodell's (1989) proposal that intrinsic motivation can be enhanced in the classroom by providing curiosity, challenge, fantasy and control. *Curiosity* is triggered by novel or discrepant events (such as the cross-shaped shadow) that stimulate interest thereby providing the drive to resolve inconsistencies between their intuitive explanation and observations. *Challenge* concerns that level of difficulty that is sufficient to allow the student to experience a sense of mastery and achievement when they are successful in the task; this is shown above where the student describes the 'dawning of understanding'. The difficulty for the teacher in this sense is to match the *appropriate* degree of challenge to student ability and level of understanding of the subject, and it is likely to be one reason that introducing cognitive conflict produces a variable result in practice. *Fantasies* allow students to step outside of and to make comparisons with real life, and the scenarios we describe here permitted this in the sense that they provide experiences that are uncommon in real life and require individuals to make comparisons with everyday experience of shadows. Lepper and Hodell's notion of *control* concerns students' feelings of self-determination and autonomy in that motivation is likely to be increased when students feel involved and in control of their learning. Metacognitive approaches in learning contribute much to student control of learning and are known to facilitate the process of conceptual change (Beeth and Hewson 1999). While a review of achievement goal theory is not possible here, students are known to have multiple goals in learning including *social goals* associated with gaining approval of teachers and peers, *performance goals* relating to their ability to secure a good grade and *mastery goals* that pertain to mastery of the work (Ames 1992; Kaplan and Maehr 1999; Pintrich 2000; Xiang et al. 2003; Deemer 2004; Palmer 2005). Such research suggests that mastery goals are the most beneficial when it comes to learning subject and these are related to the use of deep learning strategies. Practices that promote meaningfulness, involvement, collaboration, participation in shared decision-making and contain regular feedback and praise for achievement contribute towards students' mastery goals.

The metacognitive approach underpinning pre-service teachers' learning contributes towards their confidence to engage critically and tolerate uncertainty. To be effective in practice, such an approach must form a central tenet of course philosophy if it is to raise awareness of professional implications for practice. Its success is demonstrated in students recognising that the varying degrees of cognitive conflict had resulted in a differential response within the group (Fig. 4.6a. iv). Cognitive conflict with a low degree of disturbance had failed to engage thinking further, whereas the cross-shaped shadow produced a profound effect that had impact on, not only on their drive to resolve the dilemma, but also on their willingness to question the phenomenon further (Fig. 4.6a.v). The tension between disturbing thinking productively to stimulate conceptual development and disturbing thinking to such an extent that a learning barrier is created formed an important point for debate within the group:

> Care should be taken not to demoralise and so demotivate the learner ... I realise that I need to be tactful and support children in my care, nurturing them in order to maintain or build up their self-confidence.

The fine line between motivation and demotivation can only be discerned and managed when the teacher proceeds with caution and sensitivity in making professional judgements about how to support learners appropriately. The implication arising from this is that teaching must address the learners' agenda and that it must provide space for learners to articulate and examine their ideas. Students recognised that the learner must be supported adequately through the resolution process and presented with scenarios in contexts that they can relate to if cognitive conflict is to provide a positive force in promoting understanding in science.

4.5.2 Pedagogy Relating to Light

Figure 4.6b summarises the groups' perceptions of the difficulties inherent in the teaching and learning of light and shadows. A key feature of reflections concerned the abstract and counterintuitive nature of the subject itself and, in particular, the problem of locating cause and effect in what appears as an instantaneous manifestation (Fig. 4.6b. i,ii):

> Light is a very difficult topic, the problem is that you can't see it happening. You can't look at an image and see how it's formed; it can only be explained by diagrams. We see what it does not how it does it.

A fundamental problem in teaching simple optics resides in the need for the learner to define *what light is* and, for teachers to recognise the problems presented by this in learning. In identifying the difficulty presented by their own need for qualitative definition of light, and the description the behaviour of light provided by science, students began to consider some implications for teaching and learning. For instance, some discussed the need to 'freeze frame ' the process mentally in order to understand it and to isolate light production, propagation, interaction with materials and shadow formation, even at a simple level:

> The fact that light 'actually travels' is an area of potential difficulty. The difficulty lies in passing the knowledge onto children in a way they can observe.

Suggestions were made for structuring children's learning in this area, such as the use signals as a means of developing the notion of light travelling from a source and using the organising framework of light production, light travelling, light interaction with objects and light reception as a useful aid to planning:

> Having a structure like this allows teachers to scaffold children's learning about light. It's much easier to understand if we break it up into smaller pieces and then put them back together again.

In making such observations, a rationale is being constructed for the subject within the curriculum and the implications for the sequencing of teaching and learning are being considered spontaneously. One learner, whose initial conception was that light acts as an aide to vision and that vision is possible in the absence of light, attributed this idea to the fact that in her previous education light and vision were taught as separate topics and this led her to underestimate the role of the observer in optical systems. According to Galili and Hazon (2000), the role of the observer is a crucial and possibly understated part of learning in this domain.

The language used to describe shadow formation was also recognised as a potential learning barrier (Fig. 4.6b.v) and, in particular, notions of shadows *belonging to* objects or being the *image of* an object were identified as a major source of misconception. Also problematic were the visual representations of light travelling found in texts that depict light travelling in one direction only from a source in the form of a beam.

4.5.3 Pedagogical Implications for Future Practice

Despite being in the final year of professional education that had focused on the importance of children's ideas in learning science since the outset, this experience resulted in an almost revelatory view of the centrality of the issue, captured through narrative such as:

> I've come to realise that unless I try to really challenge children's misconceptions in my teaching, they may not change their minds about what they think is happening, it's no good just telling them the right answer, it's much harder than that.

Students identified a need to develop a more in-depth understanding of not only subject but also curriculum requirements, children's conceptions, strategies for promoting conceptual change and how learning might be scaffolded in the domains they would have to address in practice (Fig. 4.6c.i). Moreover, they considered factors that would be important in their future planning for teaching (Fig. 4.6c.ii). Engaging and motivating children was, clearly, an important issue as was the problem of what constitutes *appropriate* explanation. Their reflections revealed a keen sense of empathy with the condition of being a learner in science and there were some indications of its effect in planning and teaching:

> In looking back on how my own ideas have had to change as a result of them being found to be incorrect through experiments I have, I feel, changed the way I perceive science learners from now on. The overconfidence I admit to having at the start of the light unit has been shaken and was not as solid as I thought … it was quite humbling and has encouraged me to better understand the roots of misconceptions children might have. The fact that light is not the only abstract subject I will teach in science serves only to increase the potential for my new approaches towards teaching science … the changes have already had an impact on the way I talk with children. When carrying out an experiment in school I found the way I talked about light with children was focused. I found myself constantly referring back to the fact that light needed a source and would ask the children to identify the source in the experiment or I would ask the group about how they thought the light was travelling.

4.6 Discussion

This chapter illustrates not only how anomalous data pertaining to shadow formation are assimilated and accommodated, but also the powerful influence of motivation, drive to find resolution, and emotional reaction in learning science. Although some researchers (Dreyfus et al. 1990; Elizabeth and Galloway 1996; Dekkers and Thijs 1998) argue that instruction based on cognitive conflict does not necessarily promote conceptual change, we have found that, if used carefully, it has much to contribute to the learning of science in the classroom. Even though the examples offered in the first two scenarios failed to engage most learners in what Limón (2001) might consider meaningful cognitive conflict, we believe that such low level conflict has value in orienting thinking and beginning to focus student attention on anomalous elements of the phenomenon. It represents the early stages of arousal of interest and the beginnings of identifying discrepant events. Scenarios 1 and 2 succeeded in focusing attention on the light source, an element given scant recognition in initial thinking about shadow formation. At this point, observations were generally rationalised within existing schema, although two individuals explicitly acknowledged that the outcome was contrary to their thinking and experience of shadows, commenting that in 'real life shadows don't look like this despite there being multiple light sources in a room'. Such marginal anomalies suggest the beginnings of engaging in cognitive conflict as observations begin to militate against existing explanation of shadow formation. It may be that this preliminary stage of orienting thinking paved the way for the deeply disturbing experience of the cross-shaped light source and the cross-shaped shadow encountered in the third scenario.

Learners in a state of cognitive conflict are known to express signs of curiosity, arousal and inner drive to solve the conflict, as well as expressions of frustration, satisfaction and ultimately contentment as they arrive at a meaningful resolution (Movshovitz-Hadar and Hadass 1990). Researchers report diverse signs of being in a state of cognitive conflict including hesitancy, tension, perplexity and confusion (Zimmerman and Blom 1983). The students in our study displayed all of these elements from time to time during teaching and learning, but they were most evident in the cross-shaped shadow activity. In such a counterintuitive scenario, there lies the potential to trigger meaningful cognitive conflict and to create the drive in learners to find resolution of the disharmony. However, there resides a tension in both experiencing and employing a cognitive conflict approach in classrooms in that the potentially negative affective dimension must be managed and controlled within limits that can be tolerated by the learner. Ultimately, this must be balanced by the positive affective dimension of succeeding in resolving the conflict. This demands considerable professional judgement on behalf of the practitioner in terms of his/her knowledge of learners' in general (including learning styles, interests, confidence, social groupings, previous learning experiences, motivational goals and ability to tolerate uncertainty) as well as knowledge of the learning process, the nature of the subject matter and how best to facilitate engagement. It requires teachers to be able to detect the signs of learners in a state of cognitive conflict and to make judgements about how best to support them so that periods of confusion and anxiety are not extensive and do not lead to demotivation.

A key element of supporting the pre-service teachers in finding resolution in regard to the cross-shaped shadow was the deconstruction of the shadow into the vertical component followed by partitioning it into parts with the ruler. This was instrumental in encouraging learners to consider the source in terms of multiple points (and ultimately infinite points) and this led to the spontaneous construction of diagrams of how light travels from a source. The conflict also served to precipitate a process of reflection, reading and discussion in which students continued to develop their metacognitive knowledge and skills. The metacognitive dimension is critical both in recognising the inadequacy of personal reasoning and addressing cognitive disturbance productively; it affords the opportunity to accept that tolerating uncertainty is a normal part of both the practice of science and the learning of the subject. The reflexive approach promotes critical scrutiny of explanations and consideration of their usefulness in learning. To this end, the physics explanation of the cross-shaped shadow provided by Feher and Rice (1988) proved remarkably inadequate because it implicitly assumes a holistic schema of light production, propagation and visual perception. In deconstructing the meaning, the learner is required to integrate these elements. More importantly, the students were able to recognise both the potential and limitations of their attempting to make sense of the phenomena.

4.7 Some Concluding Remarks

The instructional practice of cognitive conflict has played a major role in conceptual change research and teaching over the last four decades. Having been somewhat eschewed latterly in science education research as a practice that produces a variable response in learners and learning, it is difficult to conceptualise how radical conceptual change might be promoted *without* placing a learner in a situation whereby they need to consider the explanatory inadequacy of their existing explanations. Indeed, the awe and wonder we can inspire in learners by exposing them to the unexpected behaviour of the physical world ultimately resides in recognising that our expectations have been challenged. The refraction of light as it passes through a prism, the difference between lighting bulbs in series and lighting them in a parallel circuit, the floating of an iceberg or the magnetic repulsion of magnets all have the potential to inspire both teachers and their pupils. In current practices based on extending experience, introducing alternative perspectives, promoting dialogue, discourse and argumentation, the learner is invariably exposed to conflicting reasoning that may precipitate cognitive conflict. It remains a cornerstone of current teaching practice (and is indeed not exclusive to science in this sense). Teachers have always employed the strategy of helping learners to see why their ideas fail to explain outcome as an important tool in promoting thinking and learning and probably always will. Examples discussed in this chapter showed that the depth of cognitive conflict is significant in triggering *meaningful* conflict, but it may be that a more prolonged exposure to low-level conflict might have a similar effect. We would agree with the view of Baddock and Bucat (2007) and Limón (2001) that the success of cognitive conflict as a teaching strategy is dependent not just on the nature of the conflict presented to students, their ability to perceive it and their motivation and ability to resolve it. The manner in which the teacher interacts with both phenomenon as well as the students is also significant. There remains much to learn about managing conflict in effective ways in real teaching situations and we believe that the development of metacognitive awareness of the learning process can be used positively to influence learners' perceptions of cognitive conflict and to influence their responses accordingly. Not only that, it is able to provide teachers with valuable insight into the nature of their students' learning in specific domains of science.

Chapter 5
Language Interpretation and Meaning

A central concern in science pedagogy is that of addressing the issue of science making sense and having meaning for the learner. The mechanism and structure of how individuals access knowledge and construct meaning have formed the focus of research into the congruence of the learner's interpretation of phenomena with that of scientific explanations. The problem of meaning, concerned with the communication of science knowledge in science education discourse, raises the central issue of language, and this has become increasingly significant in contemporary debate within science education research (see e.g. Carlsen 2007). In particular, socio-cultural theorists' claims that learning science is a discursive process (Mercer et al. 2004) derived from the traditions of Vygotsky (1978) conceptualise language as a tool for reasoning. In this chapter, we explore this perspective in relation to the presentation of science knowledge in the science curriculum and examine the epistemological notions that both implicitly and explicitly underpin this. We offer exemplification from practice which questions the idea that through language, we can somehow access the world directly to determine what there really is 'out there'. In challenging the perception that language is a tool that is applied to developing an understanding of concepts encountered in learning science, there is a subtle shift in emphasis from the idea that reality is 'out there' as phenomena *reflected in* language, towards a notion that reality is *produced by* language; an ontology compatible with a dialogic in which *meaning* is derived in discourse between teacher and learner and through peer group interaction.

If these discourses are conceptualised as dialogic in the sense that *meaning* is generated through the juxtaposing of ideas expressed in discourse, then 'this implies that meaning cannot be grounded upon any fixed or stable identities but is the product of difference (Wegerif 2008: 349) between them. One outcome of this perspective is that it applies to the generation of meaning in relation to both science concepts and pedagogical insight. This suggests that the teacher's perception of pedagogy in relation to explanation is fluid rather than fixed and is dependent on a dialogic experienced during the process of teaching, as well as subsequent reflection on that teaching. Insight into pedagogical and subject knowledge is therefore consequent on teaching. The implications of this in researching teacher's subject and pedagogic knowledge would point towards a methodology that embraced a

D. Heywood and J. Parker, *The Pedagogy of Physical Science*, Contemporary Trends and Issues in Science Education, vol. 38, DOI 10.1007/978-1-4020-5271-2_5,

process of capturing learning as a dynamic in situ in an attempt to glean insight from the perspective of the learner. In such a model, both researcher and teacher are conceptualised as learners because there is no fixed or final interpretation of an absolute meaning that can be imposed before or after the dialogic. The mapping of 'shift' in learning is therefore less a 'journey towards' than a 'condition of being'. These perspectives are predicated on how language works, and the following account addresses some of the contemporary debate in respect of this. It is our contention that a dialogic paradigm of science learning is more commensurate with the idea of *conceiving* than with that of *conception*.

5.1 Conceptualising How Language Works

5.1.1 A Brief Look at Language as a System or Structure

In describing language as a '*tool*', the metaphor can be readily extended to include the idea of '*repairing*' *mis*conception leading towards a notion of '*fixed*' meaning that awaits linguistic expression; an appeal to some ultimate authority about what concepts mean. In examining how language works as a system or structure, this conceptualisation that language is a tool that the individual applies to making sense of experience is brought into question.

In this section, we will briefly examine language structure from a Saussurian linguistic perspective. The main work credited to him, the posthumous publication of the seminal text *Course in General Linguistics* (1966), is a result of the combined efforts of a small number of eminent scholars of Saussure who felt it important to produce a lasting record of his contribution towards understanding the structure of language. Despite being largely disregarded by contemporary writers in the field of linguistics (Brown 2001), his work is important on two counts. First, it highlights some important technical issues concerning how language works as a system because in a Saussurian paradigm, 'language is not a nomenclature that provides labels for pre-existing categories; it generates its own categories' (Cutler 2000: 59). Second, it contextualises subsequent and contemporary debate in language. It challenges the commonsense belief in a correspondence theory of knowledge in which reality is 'out there'; a fixed order of things merely reflected in language (Eagleton 1997: 93).

The central tenet of the thesis attributed to the work of Saussure concerns that of communication and interpreting meaning from *signs* and *signification* in language structure. His view that 'without language, thought is a vague, uncharted nebula' and that 'there are no pre-existing ideas, and nothing is distinct before the appearance of language' (Saussure, 1966: 111), while contentious is at once both radically disturbing and potentially illuminating. Saussure developed a structural approach to linguistic analysis in which language is objectively defined in terms of speech acts. Since speech acts require social discourse, language, which is constituted by such acts, has the dual characteristics of being both a social product and an individual construct. Language would have no meaning without individual interpretation, and communication would not be possible without socially accepted conventions that can be both accessed by the individual and communicated socially. This duality of the language system (*langue*) and the individual facility to access language conventions (*parole)* are the basis on which the linguistic system is analysed.

Saussure considered language as a system of signs to be studied synchronically as a complete system at a given moment in time, rather than diachronically as in its historical development (Eagleton 1997: 84). The latter implies a relative degree of instability of the sign; language in an evolutionary state. The historical and contemporary meanings of the word 'gay', for example, are significantly different. Synchronic analysis enables the exploration of language structures that would otherwise be distorted through diachronic temporality. He uses the analogy of chess to illustrate the underlying rules of how language works as a system. In a game of chess, the structure of the game, its rules or 'grammar' in respect of the moves for and between the pieces, the number of pieces and squares on a board are the same regardless of where the game is played; this applies equally to a game taking place in Asia or Europe. Similarly, these rules (structures) are consistent whether the pieces are large, small, or made of ivory or wood. The game therefore maintains its 'wholeness' as a system while being able to accommodate transformation within parameters that are self-regulating (Jones 1999: 64). The limits as to what is possible in any game are determined by the game rules (grammar). The only way a game could change significantly would be if there was some *internal* change to the structure of the game itself (e.g., changing the number of pieces and squares or how pieces can move). In understanding the game of chess, it is the internal structures that we need to pay attention to, and not the specific moves in a particular game.

For Saussure, the linguistic sign is entirely a mental phenomenon. The *sign* is a combination of a concept and sound image. The concept that which is *signified,* derives from the sound image (signifier); that is, on hearing a particular word (sound image), the concept is evoked. For example, the word 'tree' is the sound image signifier that evokes the signified concept tree. The concept is that which is *signified,* and the sound image is the *signifier.* The relation between signified, signifier and sign formed the fundamental framework and structure for his for linguistic analysis.

5.2 Sign and Signification

Two important principles are identified with respect to the linguistic sign, and these challenge the idea that the sign is a referent corresponding directly to a thing. First, and what is seen as of critical importance is the *arbitrary* nature of the sign that is derived from the fact that the bond between the signifier and the signified is arbitrary. The word signifier 'house', for example, in English is an arbitrary sound image linked to the concept house (an alternative sound image would do, and indeed does operate in a different language, e.g. *maison* in French). It is, therefore, axiomatic that the linguistic sign, the synthesis of the associating of the signifier and signified, is arbitrary. The relevance of this concerns the nature of social constructs of language and convention in which the individual must participate. The individual does not have unilateral, arbitrary choice of signifier since the signifier needs to be socially communicable to the linguistic community. The arbitrary nature of the signifier lies in its relationship to that which is signified. The Saussurian sign is entirely a mental phenomenon not to be confused with the referent or object itself:

> The Saussurian sign, then is an abstract object, it is not to be confused with whatever it is the sign of, with something in the world ... people assume that signifieds pre-exist signifiers, or that meanings await expression. ..... The arbitrariness of the linguistic sign is a more radical matter than is sometimes realised, because it establishes the autonomy of language in respect of reality. (Sturrock 1993: 15–16 *[our emphasis])*

This disrupts common-sense notions of how language works and challenges traditional referential models that assume fixed relations between object and meaning such as those derived from Piagetian notions (Walkerdine 1988:3) in which the relation between signifier and signified is one of representation.

The second principle concerns the idea that *significance* is determined not through the intrinsic value of the sign in isolation, but by the sign's position and relationship to other signs; for signs to have meaning, they must compose 'of a dissimilar thing that can be *exchanged for* the thing of which the value is to be determined and similar things that can be *compared with* the thing of which the value is to be determined' (Saussure, 1966: 117–120). Signs are

> purely differential and defined not by their positive content but negatively by their relations with the other terms of the system. Their most precise characteristic is in being what the others are not. ... Signs function, then, not through their intrinsic value but through their relative position. ... Everything that has been said up to this point boils down to this: in language there are only differences. (Saussure, 1966: 117–120)

This implies that 'concepts are like holes in a net: specified by their boundaries but empty in themselves' (Harland 1994: 15). The meaning of a sign is determined by its relation to other signs and is derived from either *syntagmatic* or *associative* relations. The former refers to how one sign would link together with others in a linear series (concatenations). The meaning of the word water, for example, is dependent on its position in time and discourse: waterfall/water shortage/drinking water/in deep water. Associative relations refer to those psychologically linked entities and these are indeterminate in order and indefinite in number. Thus, the word 'weight' is attributed meaning because it is a particular property of an object that is different from other properties such as shape, colour or size (in language there are only differences). It also acquires signification through metonym, in which one sign is *associated* with another and metaphor, in which the sign is *substituted* for another. In the following discussion, we explore further how signs function in relation to science learning.

5.3 Signification in Science Learning

Saussure was concerned with the internal structures of the linguistic system. The two key notions of his linguistic analysis, the arbitrariness of the linguistic sign and that the meanings of signs are purely differential, provides some insight into the problem of language and learning science. The emphasis on the sign being entirely a mental phenomenon resists any attempt to see language as merely reflecting objects and referents in the world. While structuralist approaches to understanding language in science learning are not without problems, the most evident being the almost exclusive concern with the social part of language (*langue*) and the marginalisation of individual acts of communication (*parole*), it is important to remember that the Saussurian paradigm is concerned with *how* language works rather than what it says. It does, however, offer some mediation to this constraint because signification derived from a process of difference is not static and:

> [i]n emphasising the indissolubility of the signifier and the signified in the sign, it is also necessary to stress their separation. In other words, it is the signifying chain which produces the chain of signifieds. Language then becomes a ceaseless productivity. (Coward and Ellis 1977: 23)

This 'ceaseless productivity' derives from the 'shift' in meaning consequent on dialogue referred to earlier and applies to both teacher and pupils (and researcher); a shift that raises the possibility that meaning is necessarily deferred. We provide below some exploration as to what such an analysis might look like and identify some of the limitations that point towards a more interpretative mode.

Consider the signifying chain associated with the word *weight* as it might be understood. In science it is necessary to consider weight in a quite different way to ordinary, everyday use because a scientific view of weight is that weight is a force. The associative relation of the signifier weight within a defined framework of force does not simply emerge through tactile experience. In drawing attention to the effect that the force of weight has on structures or floating objects (discussed in more detail in Chapter 2), what is signified by the word weight would encompass the idea that weight is a force, but not necessarily the consequences of this for how objects behave in water, how they fall to earth and how structures support a load. The interpretative process is never complete; it requires the constant review of words and their meaning in different contexts. This can be seen clearly in Chapter 2 where we discuss the difficulty students experience in achieving context independent conceptual change in regard to the notion that weight is a force. Each stage of the process involves comparison and contrast of the new with existing experience in order for ideas to evolve.

The evolution of meaning is a constant dynamic. For the individual to access this (*parole*), the demands are considerable because the norms and rules of science (*langue*) not only deal with abstractions, which require of the individual a reorientation of common experience, but also require that this process be undertaken using the same words:

> The difficulties are greater than we will probably ever realise: they may be illustrated by imagining that we are attempting to interpret a sound chain in a foreign language … if the language or accent is unfamiliar, then it is unlikely that we will even be able to differentiate one signifier from the next. The signifiers themselves may not be just single words but phrases: possibly ones where the whole is not equal to the sum of the parts. (McNamara 1995: 150)

The common everyday use of the word weight is unlikely to be associated with the idea that weight is a force dependent on gravitational attraction between the masses of bodies.

There is a range of interpretations from which meaning could be derived other than from the specific manner in which the term is used in science. To speak of a person losing weight usually signifies a change of shape by redistributing mass to fit with the body fashion of the time. For most people, weight loss manifests itself in an individual changing shape and this can lead to the implicit notion that smaller equals lighter.

5.3.1 Paradigm Constraints in Reasoning

As explained in more detail in Chapter 2, developing a meaningful causal mechanism to explain floating and sinking within a paradigm of forces requires holding two concepts simultaneously, weight and size. For objects that float, the downward force (the object's weight) exerted on the water by the object is balanced by the upward force (upthrust) exerted on the object by the water. The force with which the object is 'pushed back' is related to the size of the object; the bigger the object, the more the water can be pushed out (displaced); we might say. The more water is pushed out of the way, the more it pushes back. This all sounds terribly wordy and could be more succinctly summarised in terms of the relative densities of the object and fluid but this highlights the constraint of working in a particular paradigm. The idea of a relationship between weight and size is a qualitative beginning of understanding density, a key underlying concept in floating and sinking. The signifier density and its associated relations, however, do not provide an explanation that derives from causal mechanisms within a forces paradigm of reasoning.

> [T]he mass of atoms and the spacing between atoms determine the density of materials. We think of density as the "lightness" or "heaviness" of materials of the same size. It is a measure of the compactness of matter, of how much mass is squeezed into a given space; it is the amount of matter per unit volume: Density = mass/volume. (Hewitt 1989: 196)

Attempting to translate this into an interpretation of causal factors with respect to density is particularly problematic. In coming to understanding, the reference frames from which meaning has evolved are not related to the concept density in the sense that the quantitative definition of density as mass per unit volume is quite independent of force. The quantity of matter in a body, its mass, is not changed by gravitational force so both mass and volume remain constant on either the Moon or the Earth. This would apply equally to the floating object and the water resulting in no change in the floating position.

The relation between the signs here is problematic because a specific sound image, the signifier density, requires a chain of signifiers to be considered simultaneously. The synthesis of the 'parts', weight and size, is required with a changed etymology of the word weight within the conceptual framework of force in order to comprehend the 'whole' (density). This is further complicated because density is not 'defined' within the terms of an explanatory framework of forces because it is not weight per size, but mass per volume; a relational rather than substantial concept that is (or can be) without reference to forces. The idea that it is an object's weight for size which determines whether it floats or sinks is dependent on the parts (chain of signifiers), weight and size, resonating with the signified concept in which the word density is defined. It could look something like this: we have a word that describes whether an object is heavy or light for its size, the word is *density*. Density is weight for size. In attempting such encapsulations, including the more scientifically acceptable definition that density is mass per volume, there is a danger in becoming seduced by the powerful intoxication of definition. It is important to recognise that

> to consider a term as simply the union of a certain sound with a certain concept is grossly misleading. To define it in this way would isolate the term from its system: it would mean assuming that one can start from the terms and construct the system by adding them together when, on the contrary, it is from the interdependent whole that one must start and through analysis obtain its elements. (Saussure, 1966: 113 *[our emphasis]*)

So the above definition of density is a complex whole to which we can trace back the parts that constitute the signifieds. It is a quantification of that which is qualitatively explored. It illustrates the need to juxtapose between the synchronic, static and historical derivative of the signifier weight, as used in everyday language, and the diachronic evolution of the concept as characterised in the *langue* of science. Such assimilation of ideas requires a reinterpretation of the words weight and size (within e.g., a cognitive framework of forces to explain floating and sinking) and the introduction of a new signifier, density, which 'captures' the relationship between the two in terms of an object's properties. For the word density to have signification it is necessary for the associative signifier, weight, to signify force and to consider the linear syntagmatic relations of weight and size. However, the referent *force* from the word weight would require a juxtaposition of meaning between two apparently contradictory world views, the common one and the scientific one.

5.3.2 The Relational Value of the Sign

Consider the following in respect of the relational value of signs. While working with students on developing their understanding of forces (detailed in Chapter 2), a student reasoned that an object floating in a tank of water would float in a higher position on the Moon because the object would weigh less as a consequence of reduced gravitational attraction. The interpretation focused on the loss of weight (due to the decrease in the pull of gravity by the Moon compared to the Earth) of the object. The object would therefore not push out as much water and would therefore float in a higher position. The plausibility of this line of thought is understandable and probably derives from the tactile experience of pushing a balloon in water and relating this to investigating the forces acting on the jar. For the physicist, it misses the crucial point that the floating position of the jar is not simply determined by its properties of weight and size but is consequent on the *interaction* between the object and fluid.

In addressing this idea of interaction, the relational meaning of the sign is problematic. The force exerted by the fluid is referred to as *upthrust*; in tactile terms it pushes *upwards* against the *downward* force exerted by the object (as in the experience of pushing the balloon in water). It could be argued that the meaning attributed to the sign upthrust is consequent on it being *different* to downward (or sideways or tangential and so on). In the thought experiment referred to by the student, the upthrust exerted by the fluid on the jar would be proportionately reduced with that of the decreased downward force exerted by the jar on the fluid. This is because the upthrust that a fluid exerts is consequent on gravitational force. To perceive of an upward force as being dependent upon and proportional to the fluid's weight (a 'downward', i.e. Earth-centred force dependent on gravitational attraction) seems entirely incongruous with signification from the signified *up*thrust. This requires a radical shift in the signs semantic relation in terms weight and force. That is, conceptualising the force of upthrust as being related to the weight of the fluid in a quite different way; as a property of the structure of the fluid, an attribute that enables the fluid to 'collapse' around an object positioned within its structure. Downwards and upwards are important root derivatives here. That is, weight as a downward (or more precisely Earth-, or in this instance, Moon-centred) gravitational force exerted on the mass of the object, and the opposing paired force upwards as upthrust exerted by the fluid on the object.

The student interpretation of the word upthrust, from the scenario on the Moon, is that upthrust is an 'upward' force in direct opposition to the 'downward' force (weight of the jar). This was probably a consequence of a practical investigation into changing the weight for size ratio of the jar in order to explore more fully the forces acting on a floating object. Adding more water into the jar, for example, increased the downward force (weight) of the jar. This increased the weight per unit size (mathematically expressed as weight/size) and, consequently, the jar floated in a lower position in the water. In order to balance the additional downward force, the upward (or opposing paired force) of the water must also increase and this can only happen if a greater volume of water is pushed out (displaced). On the Moon, where the gravitational attraction is considerably less than it is on Earth, the student reasoned that the weight of the jar would be less. Her hypothesis was that this reduced weight for size (since the volume of the jar would remain unaffected) would mean that the downward force would be reduced and she therefore predicted that the jar would float in a higher position in the water. The play on words now becomes increasingly problematic. The following account (Rowlands 2006: 8) presents an explanation within the conceptual framework of forces that offers some insight in how gravitational attraction influences the behaviour of a fluid.

	B	**A**	**C**	
	a	**x**	**c**	
	b	**y**	**d**	

'Think of a cube of water A. In effect it is 'floating' – at least it is not falling to the bottom! What are the forces involved in keeping it up? It can be thought of as 'resting' on the top of the column of water xy. Water A presses down on the column which must therefore be pressing back upon A. So the question becomes – what supports the column xy? By itself the column would simply collapse under the weight of A. The column must therefore be supported on either side, that is, by columns ab and cd. How do these columns manage to press sideways on xy and thereby support it and prevent it from collapsing? It is because they have the weight of cubes of water resting on them, that is, B on ab and C on cd, tending to make them collapse outwards. Their tendency to collapse outwards is balanced by the tendency of xy to collapse outwards! (Columns ab and cd are balanced on their outsides by other columns of water, and so on until the sides of the container provide support for the water.) The two tendencies are balanced because the weights of A, B and C are equal.

Really, this argument needs to be stated in terms of water pressure rather than columns of water – but the result is the same. It is, though, more difficult to visualise since you have to think of 'very small cubes' – points of water But there is a kind of 'pressure field' which is equal on points horizontal to each other but which increases with increasing depth of water.

Now think of an object floating in the water, displacing the cube of water A:

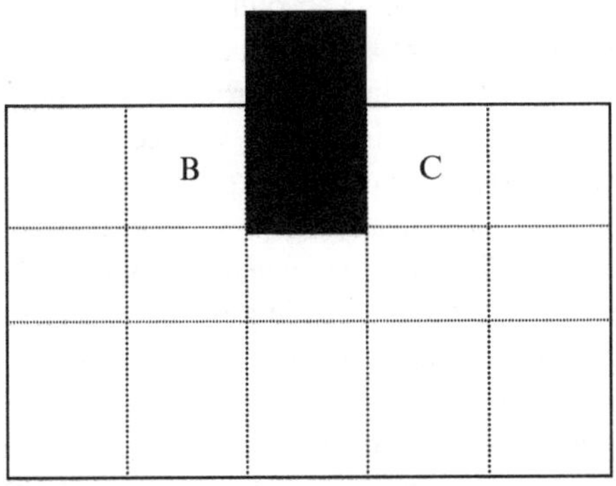

For the object to float, its weight must be equal to the weight of water displaced. Any other arrangement would not be in balance. If the weight of the object were to be more than the weight of the water displaced, then B, C, etc. would not be able to balance the weight of the object, which would therefore sink further down. Of course, if the object when entirely submerged did not displace more than its own weight of water, it would continue to sink. On the other hand, if the weight of the object were to be less than the weight of the displaced water, the forces would be unbalanced and the object would be pushed upwards.

Focusing on the weight of the object and the weight of the displaced water is misleading – particularly if you think of an object floating in only a small amount of water. How does the object below manage to float?'

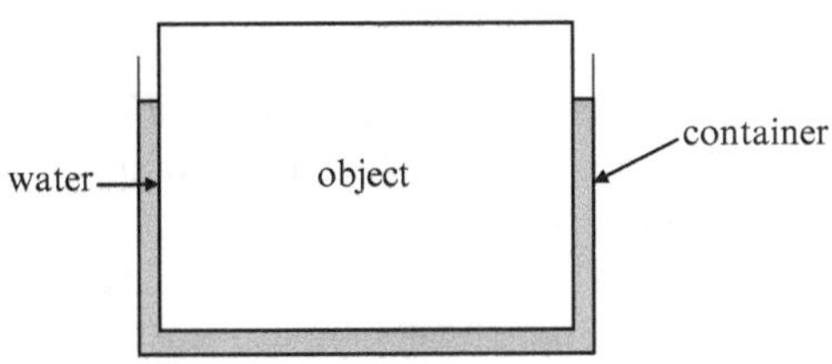

> As mention above, the explanation should be put in terms of pressure rather than total force. As long as pressures are in balance, the object floats. Thinking in this way points to the interesting idea that there is not discontinuity in the pattern of the 'pressure field' across the boundary between the object and the water. In other words, the pattern of pressure is the same across the water and object as it is in a container of water alone (and the same as it would be in a solid object with the same density of water).'

In this account the understanding of upthrust and its relationship to floating objects is being conceptualised in terms of the structure of water as a fluid material. The 'supports' and 'columns' could be thought of in terms of a 'collapsible' property of the fluid whereby the structure of the material 'collapses in on itself' as a result of its own weight. Another way of looking at this would be to say that the material is too heavy for its strength. This property of 'collapsing' could offer an explanation in terms of the student's reasoning; that is, the loss in weight of the jar would be balanced by an equal loss in weight of the water that would reduce the 'collapse force' of the fluid in which the upthrust would be relatively less. The description offered is a re-reading of the word upthrust that offers some way of resolving the contradiction of an upward force being dependent on a downward force! Any analysis of the word 'upthrust' in terms of an absolute definition is of course, as Saussure reminds us, an ineffective procedure. Although a structured approach to the problem of meaning and language can be seductive, particularly if the explanation resonates with personal reasoning, any linguistic analysis needs to be tempered with the acknowledgment of limitation in relation to explanations of how things are. This is because to seize the text synchronically as an' object in space rather than a movement in time' (Eagleton 1997: 116 *[our emphasis]*) ignores the transient nature of the sign in relation to the individual's reading of the text of science and matching this to life experience. In summary Saussure's model offers insight into *how* language works as a system. The arbitrary nature of the sign has the radical implication that *meaning* is produced by rather than reflected in language. In this model there is no possibility of directly accessing objects and referents in the world because any attempt to do so is always mediated through language. Since we can never access the world directly, the significance of a dialogical approach to learning and teaching science in which there is a focus on developing meaning through discourse would seem all the more important. This would involve a shift in emphasis from structure to interpretation. In the following section, we switch from a concern with *how* language works as a system or structure towards individual interpretation.

5.4 Interpretation and Meaning

A structuralist approach is concerned with how meaning is derived from a system of signs and one criticism of this focus on language as a system is that the process in which individuals construct meaning through the act of interpretation is somewhat marginalised. In the following discussion, we address this issue through revisiting the discipline of hermeneutics which was introduced in Chapter 3 when we applied it to the use of analogies in learning about simple electric circuits. We are not suggesting here that hermeneutics is in any sense related to structuralism and recognise that the paradigms would probably be seen by most as incommensurate, having a diverse historical evolution with radically different concerns. What we do propose is that each in its own way can offer a broadening of the conceptualisation of language in learning science in which the former provides insight into how language works whilst the latter enables us to better understand the interpretive process and the possibilities and constraints that language holds for this.

The etymology of hermeneutics, the science of interpretation, derives from Hermes the Greek messenger of the gods, hence the focus on the message or meaning. The discipline has its origins in theology where it developed rules and principles (canons) through the process of interpreting meaning (God's message) in biblical texts. Since then it has broadened in scope to include the social sciences more generally and a range of different approaches to the discipline has emerged. There is no firm agreement about the distinct nature of the categories attributed to these various approaches but the common philosophical position adopted by hermeneutics emphasises 'understanding as a situated event in terms of individuals and their situations' (Heywood and Stronach 2005: 115). Contemporary hermeneutics allows for a range of possible interpretations and implies that no one interpretation can ultimately be decided upon; a condition in which meaning is deferred because there always remains the possibility of an alternative interpretation. This idea rests somewhat uneasily with a correspondence theory of truth sometimes attributed to the natural sciences where there is an attempt to decide between competing theories on the basis of objectively collated empirical evidence. Despite these tensions, it is our contention that the discipline offers an alternative and interesting insight challenging some of the taken-for-granted assumptions concerning the conceptualisation of language as a tool and its associated metaphorical implications which were alluded to at the beginning of this chapter.

5.4.1 *What Counts for Text?*

Hermeneutics has certain canons that apply to the interpretation of a text. Prior to providing exemplification of these from science learning it is necessary to briefly outline the way in which the term text is being used and how meaning relates to this. This issue has been briefly referred to in Chapter 3. The case for the centrality of meaning in science education, as alluded to previously, is convincingly argued by Eger (1992a) who raises the question as to the appropriation of hermeneutics to the study of science. He goes on to argue (1992a, 1993) for such an approach but makes an important distinction between the scientist studying natural phenomena and the science student studying science in educational settings. The former presents significant philosophical difficulties which while not insurmountable, are marginal to our central concern here (see Eger 2006 for a more detailed treatment). For the purposes of this discussion, the differentiation is important because in science education, the student is confronted with an already defined, often counterintuitive world represented in science texts which require interpretation by both teacher and student. The concern with meaning of this multifaceted text, manifest in textbooks, curriculum syllabi, discourse in pedagogy and occasionally, in school science at least, science education research, firmly situates learning and teaching as commensurate with a hermeneutic act of interpretation. The question then arises as to how this might inform pedagogical approaches to science and language. In addressing this question we offer exemplification from two conceptual domains that raise both contentious and challenging issues for science education.

5.4.2 Language and Accessing the World (Electricity)

In Chapter 3 we identified some of the pedagogical implications related to learning and teaching about simple electric circuits through analogy. The attempt to access how simple electric circuits work through reasoning in a base domain in order to understand a target phenomenon illustrates some of the profound difficulties experienced in trying to access the world directly through the language we use to describe it. The hermeneutic principles of *possibility* and *constraint* illustrate the 'double-edged sword' that language presents for us in this regard. Clearly there is a need to project from where we are at in terms of attempting to understand phenomena that are not directly accessible. Such projections are, in hermeneutic terms, the fore-conceptions (prejudices, i.e. pre-judgements) that we bring to the task of interpretation. In hermeneutics, pre-judgment is a necessary condition of the interpretive process; prejudice is therefore conceptualised positively. These pre-judgements form the particular vantage point from which we project meaning and limit the range of *possibilities* that we are able to see or perceive; they set the *horizon* of our understanding of vision. The concept of horizon characterises 'the way in which thought is tied to its finite determinacy, and the way one's range of vision is gradually expanded' (Gadamer 1993: 302). It is related to having a notion about the relative significance of something that we are not quite grasping but have some sense of; a tuning in and preparedness for engaging with the text because we have awareness that there is something else there that needs to be addressed in interpreting the text. A closer look at the ways in which language mediates attempts at converging towards an absolute meaning and the application of these principles of hermeneutics can be illustrated in how an analogical reading of events relates to a literal text.

5.4.3 Possibilities and Constraints

What does analogical reasoning allow us to do? What are the possibilities and constraints evident in the process? In addressing these questions, we will look more closely at the tunnel analogy used in Chapter 3 as an explanation for a causal mechanism that accounts for the lighting of a single bulb in a simple circuit. The analogy opens up the *possibility* of explaining the lighting of the bulb in the circuit in terms of a 'friction' model of resistance. This is because the bulb in the circuit is conceptualised as a 'tunnel' constricting the space through which the electrons pass interrupting their smooth flow around the circuit. The tunnel becomes increasingly hot as a consequence of the electrons colliding with its sides generating enough heat (because of the constricted space) to make a bulb glow. This is a variation on the 'moving crowds' metaphor with all the attendant anthropomorphic attributes in which electrons become arm-waving people trying to get through the restricted space of the tunnel. In this model, energy is transferred to the sides of the tunnel via the arm waving which causes the tunnel to heat up sufficiently for the bulb to light. The electrons then need to come back to the battery to be 'recharged' to set their arms waving again. Thus, there is a 'difference' in arm waving on each side of the tunnel one side of the tunnel or 'across the bulb'. This analogy promotes a 'frictional explanatory model' of resistance.

In terms of coherence in applying the 'tunnel' analogy for a frictional (heat-generating) model, the difference in brightness for the various bulbs is explained in terms of the 'tunnel' thickness (or length). In this base domain of reasoning, we shall consider two possible contradictory explanations, one which initially appears more intrinsically coherent than the other. The first is based on the idea that the bulb will burn more brightly if the tunnel is narrower (or longer) because this would offer greater *resistance* to the 'movement' of the electrons creating a condition in which more heat and light are generated as they are 'pushed through'. In this way, the analogy defines the parameters in which we are able to reason about the problem and in conditioning thinking in this way, misses a second possible alternative explanation in which the reduced thickness of the 'tunnel' (wire) would result in less current flowing and therefore fewer electrons (arm-waving people) to cause friction. In this version, the opposite occurs, the effect being that the bulb glows less brightly. This alternative explanation is the one that holds up to empirical evidence and highlights another dimension to reasoning requiring the holding of three concepts simultaneously (current, potential difference and resistance) and explicitly recognising their *inter*dependence. The difficulties of this are compounded by the analogy constraining the capacity to think in terms of a holistic view of a circuit; the language allows us the possibility of accessing the intangible but at the same time constrains that access. The point here is that meaning is quite literally dependent on the constant juxtaposition between the *possibilities* that language offers us and the *constraints* that language imposes on us; the 'way we divide the world in language tells us how we think the world is really put together' (Gregory 1988). It is important to remind ourselves that neither of the alternative explanations derived from reasoning in the base domain and then projecting towards the target domain tells us how things really are. In respect of this, one might reasonably ask why, since they always break down under critical scrutiny, we don't simply dispense with analogies altogether and just refer to an authoritative text that will tell us precisely how things are.

This appeal to an authoritative text is particularly seductive, not least because it suggests the possibility of a *literal interpretation* as to how the circuit really works. In considering the meaning of the account below of a physicist's view of an electric circuit (Black and Harlen 1993), there is a shifting horizon consequent on the experience of engaging with the analogies referred to previously that we bring to the interpretation. The process of interpretation involves a backwards and forward movement between this 'literal' text and the analogy. In an attempt to understand the whole we break the 'text' down into parts and subsequently interpret each of these parts in respect of the whole. This circularity of movement, from whole to parts and from parts to whole, known as the *hermeneutic circle*, is a central tenet of hermeneutic method. It is concerned with convergence, attempting with each interpretation to get closer to the truth and whilst it is never possible to get there because language never allows us to arrive at the object directly, it is necessary to 'fix ones understanding in a particular explanation for the time being' on the understanding that such fixity is always contingent (Brown 2001: 36).

> Briefly, a physicist's view of electric circuits runs as follows. Charge does not pile up in conducting circuits, so current is conserved. Potential difference (p.d.), which is a measure of the field acting, is apportional around items in a series circuit according to their resistance. Free energy is transformed at a rate determined by the product 'current times p.d.' If one asks how the energy is transferred from (say) a dry cell to a small bulb, the best answer is that it is transferred through the electric and magnetic fields that surround the wire when a current flows: these fields are the means by which parts of the circuit wiring remote from the cell terminals 'know' that the current is to flow when the switch is closed: the effect of closing a switch is seen almost instantaneously (in fact it propagates at the speed of light) and does not travel around by means of shunting collisions of the electrons in the wires. The electrons travel quite slowly, and in an AC mains circuit they jiggle backwards and forwards with amplitudes of less than a millimetre. Thus no material substance enters one's house along the mains supply cable; the flux of surrounding fields delivers the energy. The resistivities of different substances cover an enormous range with ratios of the order of 10^{20} between the extremes. Because of this, a situation in which the terminals of a cell are connected by air can be regarded as qualitatively different from one where they are connected by copper - unless the air path is only a fraction of a millimetre wide, when sparks might start to occur. There need not be closed circuits for current to flow: in lighting, charges pile up between Earth and cloud by a slow pumping process until the field across the air is too great, then they flow back in a flash; the 'source' has to be pumped up again before a further discharge. Early nuclear accelerators used mains supply to charge up sets of capacitors to produce intermittent discharge in this way. However, such systems are rare, and we usually deal with closed circuits which cease to operate if the circuit is broken. (Black and Harlen 1993: 221)

This text disturbs the 'fixity' of understanding because it demands we engage with the parts such as 'free energy' and 'magnetic fields' in order to explain the whole; the parts themselves, however, can only be explained in relation to this whole. As mentioned previously, in attempting convergence as to the absolute meaning of the text, it is tempting to dispense with analogies and metaphors on the grounds that they are linguistic overlays that filter and obscure the way in which we might perceive the 'essence' of the phenomenon itself. Here we meet again the constraints that language imposes on us because language always mediates our attempts at direct experience with the world. The 'authoritative text' above is no different in this regard because there is evident an implicit appeal to metaphor and analogy in reference to terms such as 'free energy' or 'electrical and magnetic fields' and the idea that 'fields are the means by which parts of the circuit wiring remote from the cell terminals 'know' that the current is to flow when the switch is closed' serves to remind us that language defines the ontological landscape.

A possible solution to these difficulties would be an appeal to the author as appointed interpreter as to what the text really means but this is a rather naïve concept in regard to meaning because authorial intention, if it could be determined as to what was meant at the time of writing, has no privilege in regard to meaning of the text. This is because in order for any text to have meaning, it requires a reading by another. The author has no privilege in this regard. These possibilities and constraints in the interpretive process lead to the realisation that in accessing the world *we have language*, but *language also has us*. In the next section, we look at how language not only mediates but also shapes our perception of the world.

5.4.4 Shaping the Ontological Landscape

This account is based on a study (Heywood 2005) of pre-service teachers' learning about light through exploring simple optical situations. It focuses explicitly on how learners both interpreted and represented their thinking about light in practical investigative work during university taught sessions.

Vision is a dominant sense in humans and as we have seen in Chapter 4, unlike scientific explanation of other physical systems, in coming to understand light the learner is required to recognise that the observer constitutes an integral part of the optical system. It is intuitive to spontaneously explain phenomena in terms of cause and effect where ideas are guided by common sense experience and this is particularly problematic in understanding optics. Although they often recognise that light is necessary for both creation and observation of an image, learners usually cannot provide an explicit account for these processes in terms of the behaviour of light (Goldberg and McDermott 1987, Bendall et al. 1993). There have been some attempts to provide a cognitive taxonomy in relation to learners' ideas about causal mechanisms that explain optical phenomenon. These include 'facets of knowledge' (Minstrell 1992) which are defined as elementary units that ordinarily have limited explanatory powers across contexts. It is possible, however, for these individual facets to be ascribed to 'schemes of knowledge' that constitute a higher organisational level representing a common causal mechanism shared by a group of facets. For example, the 'Image Holistic Scheme' in which an image is seen as a corporeal replication of an object that can travel, remain stationary, or turn as a whole (Galili 1996; Galili and Hazon 2000). In the context of the plane mirror, such cognitive schemas often conceive of the image being projected to the mirror, and view this as staying there. The difficulties encountered in making sense of formal representation of such optical phenomenon as an image in a plane mirror have been selected here because they provide opportunity to explore further the way in which language defines the ontological landscape.

The formal representation of image formation through diagrammatic representation using light rays provide a potentially useful explanatory tool with which to describe and explain simple optical phenomena such as image formation. However, it is well-known that students frequently exhibit great difficulty in applying them accurately (Goldberg and McDermott 1986, 1987; Galili et al. 1993). As Galili (1996) points out, learners often interpret light rays in a literal sense attributing to them a material entity (like water particles comprising water), a perception that whilst disturbing to the physicist, offers some interesting insight into how language operates in our accessing the virtual world. We shall return to this point later.

In the empirical study referred to here, learners were engaged in exploring two fundamental optical phenomena; namely, how an object is seen and how an image is formed in a plane mirror. Vision was chosen as an area of study on the basis of its fundamental importance in understanding optical phenomena. In no other area of physics does the observer play such an integral role within the phenomenon being explored. In considering vision, it was anticipated that students would necessarily engage in thinking about light propagation. Image formation in a plane mirror is a more complex phenomenon requiring knowledge of specular reflection and virtual image formation.

There were three distinct aspects to the study:

- identification of personal *interpretations* and explanations of diagrammatic representations of how an object is seen.
- identification of personal *representations* and explanations of how an object is seen and how an image is formed in a plane mirror.
- application of reasoning in increasingly more complex optical contexts.

In the first of these, the key elements of scientific explanation of the optical phenomenon of how we see an object involves the learner recognising that darkness is the absence of light, light is emitted by a source and that light is necessary for an object to be seen. In exploring the notion that light is necessary for an object to be seen, propagation of light, light reflection and light perception by the eye would be necessary constructs. More advanced explanations of the phenomena would entail increasingly sophisticated views of light propagation in terms of rays or light flux, non-specular reflection and the role of the eye and the brain in terms of perceiving light and interpreting this as an image. The purpose of this element of the study was to focus on student *interpretation* of representations.

The second aspect of the study, how an image is seen in a plane mirror, presents for learners a more complicated optical phenomenon, requiring consideration of light propagation, light reflection and image formation. It is well-known that learners find this a difficult and confusing area (Goldberg and McDermott 1986; Bendall et al. 1993; Galili 1996; Langley et al. 1997). Understanding the formation of a virtual image in a plane mirror usually entails representation of the image in terms of point-to-point mapping of the light reflected by the object to the mirror where it is reflected specularly and perceived by the eye, itself an optical device. The image is interpreted by the brain as emerging from a unique, *virtual* position in the mirror. This virtual position in the mirror was explored further in the third element of the study which addressed the more complex scenario of the observer shifting their viewing point.

Initially we will refer to the second scenario of students accounting for the formation of a virtual image (in this case of a tree) in a plane mirror. This is because it required the students to consider most of the key conceptual elements of the vision process and provided a platform from which to develop these ideas further in applying them to the more complex optical situation explored in the third investigation. In explaining image formation of the tree in the mirror, most students appeared to be operating on the premise that light travelling from the tree is reflected at the mirror into the eye. The student diagrammatic representations were sometimes described in terms of light travelling:

> The light from the tree is reflected onto the mirror, the mirror reflects it into our eyes.

While at other times, they seemed to deploy a dual representation incorporating both light and the vision process, although this was not explicitly differentiated either in terms of the observer or in respect of the eye as an optical instrument:

> The rays of light make the object reflect into the mirror; our eyes look into the mirror and see it.

Light rays were represented variously as single and parallel lines (Fig. 5.1a and b) and as convergent/divergent lines (Fig. 5.1c). Figure 5.1d shows a representation that implies a relationship between the angle of incidence and the angle of reflection. Although the majority of students referred to light travelling in various ways, reference to the image was variously described as *being reflected through (or) by* the mirror *into (or) being picked up by* the eyes. No one articulated a mechanism to explain how an image would be formed.

The purpose of the third scenario was to explore notions of the image being independent of the observer's position and to address the paradox that presents itself as a particular difficulty in the relationship of the observer (subject) to that

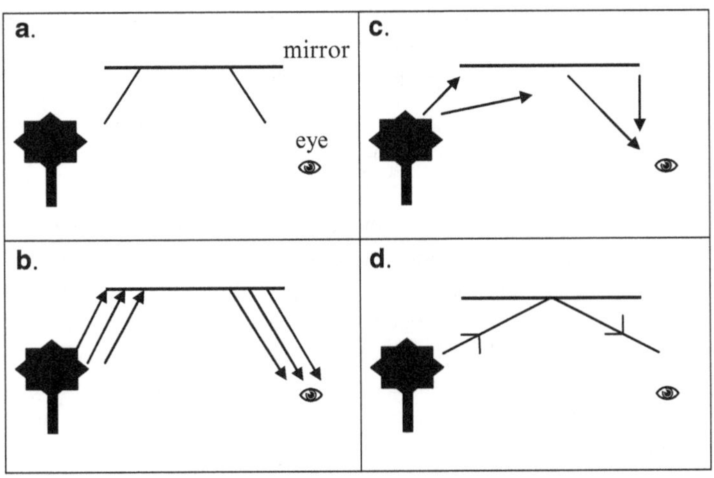

Fig. 5.1 Student diagrammatic representations of *How we see a tree* (numbers of responses in brackets, N = 36)

which is observed (object). The students were asked to observe the image of an object in a plane mirror from various vantage points. As stated previously, in classical optics, the incorporation of the observer into the conceptual framework is a necessary condition of interpretation and representation. The *apparent* movement of the image in the mirror is a subjective experience of vision; the image does appear to move, in the same way that an object appears to move relative to the background for a moving observer. In the latter case, this idea is more readily acknowledged as a relative experience (telegraph poles do not move!). However, in terms of the image in a plane mirror appearing to move as the observer moves, we meet a particular difficulty probably borne of the notion that the image is being conceptualised as a holistic independent entity. The students struggled to articulate image formation in terms of vision, light travelling and the combination of these predicates in forming a mechanism that explains what is seen. Some thought that the *line of sight* changes, not the object and its image, implying a holistic view of the image as an independent entity that can be viewed from different directions:

> It's moved but you're just looking at it differently.
> The reflection is still in the same place but because we have moved we have a different view of it.
> Others recognize the significance of the object and the mirror remaining in the same place:
> The distance does not alter but we failed to realize the significance of the block *[object]* and the mirror remaining the same.
> Because the actual observer is moving, the position of the object in the mirror moves with it.

A particular area of difficulty emerged in that there appeared to be some confusion over the notion of light travelling in straight lines and the object being reflected in a straight line. The experience prompted some to think more carefully about what it was they were seeing in the mirror; they began to consider what a reflection actually is:

> You think you are seeing the block (object) through the mirror, but in fact you are seeing the reflection. Light reflects off the block and hits the mirror and then hits our eyes.
> The light is reflected off the block into the mirror and at an angle, then it is reflected back at the same angle and into the eye and shows the eye where the image is.
> I thought that the image wouldn't move but now realise it is because the light is travelling in straight lines and <u>in all directions</u> and the eye is <u>assuming</u> that we are seeing the image. *[author emphasis]*

The explicit recognition that the image is an interpretation by the observer seemed an insurmountable problem. Unlike other areas of the physical sciences, that whilst difficult are to an extent accessible either through tactile experience (e.g. forces), or concrete frames of reference (e.g. the use of analogy in electricity), the base domain of reasoning in light seems obscure and intangible. The degree to which these difficulties can be resolved seems limited and reasoning appears particularly dependent on an appeal to an elusive causal mechanism of explanation when different contexts are experienced. There are indications of a sequential progression in cognitive reasoning although there is less evidence of a sequential progression in instruction productive in supporting the conceptual challenges encountered. In the former, moving from a holistic image projection schema, in which the image is conceptualised in terms of an entire entity moving across space, towards an image projection conceptualisation (IPC) could be considered more sophisticated. This is because in the IPC model (Galili and Hazon 2000), the student appears to understand the rectilinear propagation of light and its role in image formation, the image being an assembly of corresponding points on the object. This does provide a framework in which the phenomena can be considered as a real entity that travels through space carrying certain information about what is seen (object or image). It is based on a naïve premise that in order for an image to be formed 'it' (light) must be carrying something: the image of the object to the screen (or mirror).

In this scheme light rays are viewed literally, a single ray being considered as containing microscopic information that contributes to the image construction; a cognitive framework in which *image formation* is conceptualised as being separate from *image observation.* The difference between this and the formal conceptualisation point-to-point mapping by means of light fluxes is both subtle and sophisticated. The problem is illustrative of the need for an appeal to tangible mechanisms of explanation, a central tenet of which concerns the representation and interpretation of light rays. In this paradigm, because light rays are *represented* as explaining the behaviour of light in relation to the vision process, the rays in the diagrams implicitly and explicitly assume an ontological significance conceptualised *as light.* The language we use to *interpret* these representations, to articulate them in dialogue for explanation in pedagogy then becomes a central concern because only through this process can they take on meaning.

5.4.5 Distancing

In hermeneutic interpretation there is a need for a certain distancing in order to recognise that something would need interpreting. The fact that the image in a plane mirror is so familiar in our everyday experience in one sense precludes the need to explain it. This could be one possible explanation as to why the students in the study found it difficult to articulate their observations. Equally, students' tendency to reify non-material concepts such as images and shadows are further examples of the need to conceptualise phenomena as matter-based (Galili 1996). The implications for this latter point raise the importance of developing students' awareness of the nature of science knowledge in which experts maintain 'two distinct ontologies, one for matter based concepts (such as 'water' or in our case 'light'), and the other, for process-based concepts of a descriptive nature' (Galili 1996: 864). This is exemplified in Guesne's (1985: 29) account of the virtual image:

> The physicist interprets the virtual image of an object in a mirror by saying that the light issuing from the object reaches the eye of the observer after having been deflected by the mirror, *exactly as if* it came in a straight line from an object which would be symmetrical to the real object in relation to the surface of the mirror.

The key issue here seems to be the phrase '*exactly as if it came in a straight line*'; not that it has. The use of the word virtual here explicitly locates the ontological landscape in the language we use to describe observed events. It then becomes more than mere semantics to deliberate whether the image exists outside the subjective interpretation by the observer. In this chapter, we have presented different perspectives on language in science education discourse. We initially presented a view as to how language works as a system and followed this with an account of the way in which language conditions and shapes the world. In this latter part, it is important to distinguish between language in use and language taken as an object. Gallagher (114) refers to the metaphor of language as a window that without bringing attention to itself frames the way in which we see the world. The analogy with the window though, as in all analogies breaks down because 'language is not only an aperture to an already made world, but helps to constitute that world. It would be as if the window played a part in designing the objects that could be seen through it.'

In the introduction to this chapter, we claimed that a dialogic paradigm of science learning is more commensurate with the idea of conceiving than with conception. The examples of students' learning presented illustrate the subtle but central role of language in developing qualitative understanding of phenomena. The process of making sense of explanations and generating meaning illustrates both the possibilities and constraints of language in not only learning science but also generating science pedagogy.

Chapter 6
Metacognition and Developing Understanding of Simple Astronomical Events

This chapter focuses on the value of developing metacognitive awareness of learning as an integral part of science teacher education. Previous chapters have shown how adopting a metacognitive approach to teaching and learning affords the opportunity to support students in synthesising subject and pedagogical knowledge. In developing knowledge of their own cognition, students make pedagogical observations of significance for future classroom practice. In Chapter 4 we illustrated that as a result of experiencing cognitive conflict in their own learning, students identified emergent pedagogical implications ranging from knowledge of learners and learning in general, to detailed subject-specific observation relating to building understanding of light in the curriculum such as the need to enable learners to access what is an instantaneous process of light production, propagation and reception. In this chapter we explore the potential of this approach in generating subject-related pedagogical knowledge (pedagogic content knowledge) as students generate causal explanations of simple astronomical events. Pedagogic content knowledge (PCK) concerns knowledge related to the translation of subject knowledge in the act of instruction; it requires knowledge of the cognitive demand of the subject as well as knowledge of instructional practices appropriate to structuring learning including the use of metaphors, analogies and explanation. The chapter concludes by discussing the potential for the development of unique insight into the learning of subject in this area with important implications for instructional practice. First we consider the nature of metacognition and its potential to contribute towards effective teaching and learning in science.

6.1 Metacognition and Learning

6.1.1 What Is Meant by Metacognition?

Historically, metacognition was conceptualised by Flavell during the 1970s as 'knowledge cognition about phenomena' (Flavell 1979: 906), and represented in a general sense in literature as 'thinking about one's own thinking'. It concerns learners'

D. Heywood and J. Parker, *The Pedagogy of Physical Science*, Contemporary Trends and Issues in Science Education, vol. 38, DOI 10.1007/978-1-4020-5271-2_6,

awareness of their learning and knowledge of the processes by which that learning takes place. It encompasses the control or regulation of such processes that conscious, self-appraisal of learning can contribute towards. As Flavell's original contention illustrates:

> Metacognition refers, among other things, to the active monitoring and consequent regulation and orchestration of these processes in relation to cognitive objects on which they bear, usually in the serve of some concrete goal or objective. (Flavell 1976: 232)

The process has been recognised by educational psychologists for some time (Brown 1987). Shayer and Adey locate its origins in the work of Piaget (1914) and Vygotsky (1978) arguing that it:

> emphasises the value to the developing thinker of becoming conscious of their own thinking and developing and practising the technical vocabulary necessary for describing various actions. (Shayer and Adey 2002: 6)

The term lacks a precise definition in regard to the way in which it has been interpreted, explored and elaborated in the literature (Weinert 1987; Hacker 1998; Livingston 2003). This is probably due to the fact that it is a generalised conception applicable to many disciplines that has been applied to a broad spectrum of fields of study. Flavell's (1987) taxonomy of metacognition made a distinction between *metacognitive knowledge* (knowledge relating to cognitive matters) and *metacognitive experience* (conscious cognitive or affective experiences in relation to on-going learning). The former concerns knowledge of the nature of learners as cognitive organisms, including the learner's own nature as a cognitive processor (person variables), how specific information encountered effects the way in which one deals with task demands (task variables) and knowledge about strategies for accomplishing the task and achieving goals (strategy variables).

A growing interest in metacognition has emerged in educational forums in parallel with a recent focus on theories of learning and, in the case of science, conceptual change in particular (see Chapter 2 for a fuller discussion). The research reports a positive link between the adoption of metacognitive practices in instruction and cognitive outcome in a variety of activities and subject domains ranging from reading to problem solving (see Hacker et al. 1998 for a range of examples). It has been linked with successful learning and associated with intelligence (Borkowski 1985; Wang et al. 1990; Butler 1998; Sternberg 1998; Veenman et al. 2002). Educational psychology research has revealed positive effects of such practices in helping learners to understand a variety of subject-related content and processes such as problem solving, concept transfer, retention of learned information and independent, self-regulated or autonomous learning (Everson and Tobias 1998; Georghiades 2000, 2004a; Howard and McGee 2001; Veenman et al. 2002).

In science education, metacognitive approaches towards teaching and learning have also yielded positive outcomes in relation to promoting conceptual change and the development of meaningful learning (White and Mitchell 1994; Baird and Mitchell 1997; Beeth 1998a,b; Beeth and Hewson 1999; Blank 2000; Georghiades 2000, 2004a; Koch 2001; Davidowitz and Rollnick 2003; Kipnis and Hofstein 2007). In becoming aware of one's own thinking and the inconsistencies between observation and explanation, there lies the opportunity to question the plausibility and intelligibility of personal explanation. Such conscious scrutiny of thinking forms the basis of some contemporary educational initiatives such as the Project to Enhance Effective Learning (PEEL) developed in Australia (see Baird and Mitchell 1997) and the Cognitive Acceleration of Science Learning (CASE) project in the UK (Adey and Shayer 1994).

The pedagogic emphasis in such approaches concerns mapping cognition:

> through the critical revisiting of the learning process in the sense of noting points of the procedures followed, acknowledging mistakes made, identifying relationships and tracing connections between initial understanding and outcomes. (Georghiades 2004b: 371)

It is not our intention here to address the range of interpretations and meanings of metacognition (see for example, Schraw 1995; Schraw and Moshman 1995; McCormick 2003; Georghiades 2004b; White and Frederiksen 2005 for further discussion). However, models of metacognition generally involve two components: the first concerns conscious thinking about cognition and the second, control or regulation of thinking that emerges as a direct consequence of raising awareness of the nature of learning. Neither is mutually exclusive and each has a range of knowledge bases and competencies that have been identified as integral to the process (see for example, Paris and Winograd 1990; Schraw 1998; Pintrich et al. 2000; Son and Schwartz 2002). Hacker considers the capacity of learners to engage metacognitively is of central concern to teachers and researchers (Hacker 1998:13). In respect of this, the foundation of our approach to teacher education is based on practice that engages learners in the articulation of their learning and refection on the processes involved in the generation of understanding of ideas in science with consequent implications for pedagogical knowledge. In this chapter we will explore how this has led to valuable insight into the teaching and learning of basic astronomy.

6.1.2 The Relevance of Developing Metacognitive Awareness of Learning in Teacher Education

The exemplification of pre-service teachers' learning employed throughout this book has been based on the adoption of a metacognitive approach to teaching and learning. It constitutes a key element of an epistemological philosophy predicated on the tradition of reflective practice (Day 1993) in an explicit attempt to engage learners in identifying personal thinking and becoming conscious of how this is influenced as they interact *in situ* within the learning context. In contrast with the pre- and post- instruction analysis of learning, typical of many conceptual change studies, the intention here is to consider learning as a *dynamic* through application of reflective strategies designed to raise awareness of how cognition is shaped and moulded. Individuals' epistemological commitments (including both learners and tutor), their motivations, the science content subject matter, instructional contexts and practices as well the physical environment are elements that are likely to impact on interpretations of this process. Chronological documenting of thinking through the use of learning journal focuses students on identifying specific elements that contribute to their learning whilst recognising, for example:

- the existence of a range of explanation (their own, colleagues and that of science)
- the differences between conceptions
- critical moments in learning
- where observation conflicts with existing knowledge and understanding
- key ideas in building explanations
- the counterintuitive nature of some scientific ideas
- the effect of the socio-cultural environment in learning
- the affective nature of learning

In focusing on the cognitive demand of making sense of phenomena, and identifying what is significant in their learning, a discourse of science is generated between teacher and students, and within and across student peer groups, that serves to involve trainees in careful scrutiny of difficulties inherent in developing the understanding of scientific ideas. This process constitutes the *problematising* of science subject knowledge (Heywood 2007), and it simultaneously requires that teachers consider what might contribute towards or detract from the resolution of conceptual challenges. Through self-reflection students are encouraged to consider the range of interpretations generated within peer group and tutor discussion as a positive resource that enables a more critical review of their own cognition.

6.2 The Conceptual Domain of the Earth and Beyond

6.2.1 The Cognitive and Pedagogical Challenge of Developing Causal Explanations of Simple Astronomical Events

As many primary pre-service, and indeed practicing teachers, will not have encountered basic astronomical concepts in their own education, this area is likely to constitute a conceptual challenge. Summers and Mant (1995) reported a considerable difference between teachers' understanding and the scientific model of explanation for night and day, the phases of the Moon and the seasons concluding that this is an inherently difficult conceptual area for teachers. Our own research (Parker and Heywood 1998) found a similar outcome with pre-service teachers and this raises the issue of how they might best be supported in developing coherent causal explanation for the basic astronomical events to be addressed in primary education.

One reason for the evident difficulty presented by simple astronomy is the deeply counterintuitive nature of the concepts involved. In understanding the day–night cycle, for example, a fundamental problem concerns matching observed events with the scientific model of explanation for the day–night cycle. The daily observation of the Sun rising in the east, moving across the sky and setting in the west seems incommensurate with a model based on the Earth's spin on its axis once every 24 h (approx.); there is also the attendant difficulty that the 'spin' of the Earth cannot be perceived physically. This *apparent* movement of the Sun around a stationary Earth constitutes a geocentric view of the Earth that persisted for hundreds of years until the work of Copernicus and other scientists in the sixteenth century presented the radically different heliocentric view of the solar system. Similarly in attempting to explain seasonal change, a common-sense causal mechanism would be to postulate that the Earth moves closer to the Sun in summer and further away in winter, and an explanation involving a model depicting the Earth's orbit as approximately equidistant from the Sun in winter and summer appears at first sight incongruous with sensory perceptions as the effect of the Earth's axis is not immediately apparent. Consequently, the teaching of basic astronomy is unlikely to be simply a case of knowledge transfer; it requires a synthesis of knowledge including knowledge of learners, knowledge of learning and knowledge of the complexities of subject matter.

Barba and Rubba (1992) who studied the differences between experienced teachers and novice teachers in their approach to teaching and learning of basic astronomy highlight the complex interactive nature of the various types of knowledge drawn upon by experienced teachers in promoting effective learning. This raises the question as to what it is that teachers need to know in order to operate effective pedagogy. Strong science subject knowledge is not a guarantee of effective teaching of subject, and Shulman (1987) indicated that good teachers possess a detailed and subtle understanding of not only the content of subject matter, but also an in-depth knowledge of how best to represent the subject in the classroom setting (PCK). We will consider this further in relation to astronomy as teachers try to construct qualitative explanations for some familiar astronomical events they will have to teach to young children.

Basic astronomy is present in the curriculum guidelines for children in many countries (AAAS 1993; NRC 1996; DfEE/QCA 1999; MoE, NZ 2007). The programme of study relating to the Earth and beyond in the National Curriculum for England and Wales (DfEE/QCA 1999), for pupils aged 7–11 years, is focused on the Sun, Earth and Moon system and periodic change including how day and night are related to the spin of the Earth on its own axis, the Earth's annual orbit of the Sun and the Moon's approximately 28-day orbit of the Earth. Although seasonal change and phases of the Moon are not taught explicitly according to the curriculum, the topics are commonly employed in extending children's learning. Traditionally primary school curricula have included a strong focus on periodic change involving the marking of the passage of time on a daily and annual basis through events in the physical and natural environment, as well as through the celebration of cultural and religious occasions. However, curriculum reforms of recent years now require in science that children develop causal explanations for simple events and begin to understand the scale, dimension and movement of bodies that will underpin explanation of phenomena such as season and phases of the Moon.

Research has amassed a body of description of learners' conceptions across ages and cultures in this domain (see for example, Nussbaum and Novak 1976; Nussbaum 1985; Baxter 1989; Schneps 1989; Vosnaidou 1991; Lightman and Sadler 1993; Vosniadou and Brewer 1992; Samarapungavan et al. 1996; Stahly et al. 1999; Schoultz et al. 2001; Trumper 2001b, 2003; Barnett and Morran 2002; Trundle et al. 2002; Sharp and Kuerbis 2005; Spiliotopoulou-Papantoniou 2007; Cin 2007). While the ideas of primary children are well-documented, there have been fewer studies of primary teachers' knowledge and understanding in this area (Barba and Rubba 1992; Bisard et al. 1994; Atwood and Atwood 1995; Summers and Mant 1995; Parker and Heywood 1998; Abell et al. 2001; Trumper 2003). Trumper (2003) in a study of elementary teachers found that they may have more misconceptions about simple astronomy than high school students. Summers and Mant's (1995) work with primary school teachers confirms that the concepts involved are both difficult to interpret and often partially, if at all, understood. Our research supported this contention in a study of pre-service teachers' understanding of the day–night cycle, seasonal change and phases of the Moon (Parker and Heywood 1998). Furthermore, Bailey and Slater (2003) report that as a result of instruction, learners frequently integrate teaching with their deeply held alternative views resulting in an incomplete understanding of the subject. Vosnaidou and Brewer (1992) in exploring children's understanding of the day–night cycle and the shape of the Earth found that older children tend to hold synthetic models of these concepts that represent an attempt to reconcile perceived differences between initial ideas and synthetic models.

There are several reasons as to why simple astronomy presents such cognitive demand. First there is the mismatch already alluded to between scientific explanation and daily observation of natural events. Trumper (2003) comments:

> More than 20 years ago various workers began to examine these very intensively, and they have produced a growing body of evidence that throws doubt on the assumption that adults and children are post-Copernican in their notions of planet Earth in space. (Trumper 2003: 310)

Alternative conceptions abound and teaching in this area is, therefore, likely to entail radical conceptual shift and the implications are that teachers acquire insight into the mechanisms of learning that pupils adopt in making sense of abstract ideas that do not resonate with their experience and view of the world from the observations they encounter. Vosnaidou (1991) demonstrated how pupils develop a chronological progression of thinking from early naïve flat earth notions towards more abstract sophisticated scientific observations of explained events. Such an evolution of thinking is supported in cross-cultural studies (Mali and Howe 1979; Klein 1982) and is indeed to an extent seen to parallel the historical development of ideas in science in this area (McCloskey 1983; Baxter 1989). That this transition in conceptual understanding requires a radical shift in thinking is highlighted by Sharp (1996) in a study of year 6 to 11-year-old pupils' astronomical beliefs. In formulating the scientific causal mechanisms for simple events such as day and night and seasons, a learner must possess a complex array of knowledge including:

- the Earth as a sphere in space
- gravitational attraction and where people live on the Earth
- the Sun as a star
- relative movements of the Earth and Sun
- earth spins on its axis once every 24 h
- scale and dimension
- light production, propagation, perception and interaction with objects

6.2.2 *Using a Metacognitive Approach to Generating Subject and Pedagogical Knowledge*

Using a metacognitive approach towards learning science in teacher education enables teachers to engage in discussion of their own explanations as well as those of others, to become aware of how their learning is influenced through social interaction with learning activities as they attempt to solve cognitive problems generated by the subject matter (self-awareness). This knowledge allows teachers to be in control of their own learning and to understand how they might learn effectively (self-regulation or management) with positive benefits for the generation of pedagogical insight. We illustrate this process by reference to a study of 52 pre-service primary teachers undertaking a university-based teaching session on the Earth and beyond designed to help them to develop causal explanations for day and night, seasons and phases of the Moon (see Parker and Heywood 1998 for full details). The study involved postgraduate and first year undergraduate students with varying educational backgrounds (none of the trainees were science specialists). It was deliberately focused on a single university-taught session in an effort to replicate realistic course input on the topic. In addition to data drawn from the original Parker and Heywood study, we have included further reference to the empirical data underpinning the research. A key strategy employed to promote metacognitive awareness of learning involved the production of annotated diagrams indicating a causal explanation for each phenomenon (see Fig. 6.1 for an example). These diagrams served several purposes in that, not only did they provide a starting point to which the learner could refer in auditing learning, they also acted as a powerful stimulus to group discussion of causal mechanisms and provided a reference from which learning could be mapped during the taught session. They facilitated articulation of points of uncertainty, provided a vehicle for the scrutiny of the use of language and helped to develop awareness of shifts in thinking at strategic points during the session.

Having established starting points, students were then given the opportunity to explore explanations in a small group situation using lamps and globes to model relative movements of bodies. For each phenomenon (day–night cycle, seasons, phases of the Moon), the accepted scientific explanation aimed at the level of older primary children was provided in written and diagrammatic form and modelled

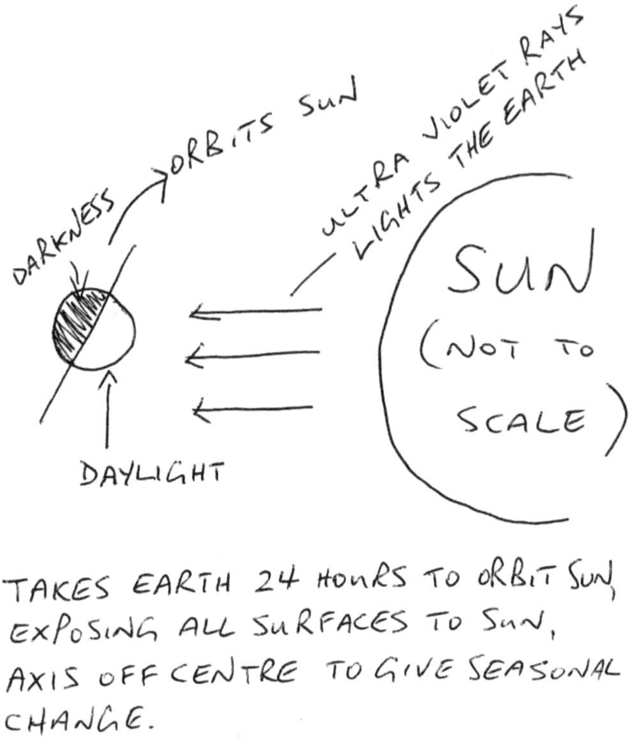

Fig. 6.1 An example of a student's annotated diagram explaining how day and night occur

alongside learners' conceptions. The tutor acted to facilitate this process through focusing student attention on similarities and disparities between explanations and their consequences, supporting students in translating ideas, providing demonstrations and helping focusing attention on what students considered to be key features of the phenomenon important in developing understanding of the concept. If thinking is to be shared in this way, professional trust must be established in the classroom such that learners feel comfortable in expressing and sharing ideas. The students in this study were used to working in a collaborative atmosphere where such practice is an inherent feature of teaching and learning. The annotated diagrams were employed simply as a mapping device; they were individual, private and used at the discretion of the student. In this process it was also important to emphasise that while some may be confident in their own causal explanations for events, others will not be able to construct what they consider to be a satisfactory explanation and in this event they will need to be encouraged to record any knowledge they deem pertinent to the phenomenon. The recognition of uncertainty is an integral part of learning science that we use as a means of enabling students to gain metcognitive control of their learning such that they are able to identify this when engaging with abstract ideas. We subsequently use the opportunity to support them in explicitly articulating the nature of their uncertainty, to generate questions and focus learning goals accordingly. This is critical in learning as without such focusing students are likely to become disheartened by their lack of comprehension and to perceive themselves as 'failed' learners of science (Parker and Spink 1997).

6.3 Mapping Movement in Conceptual Understanding About Simple Astronomical Events

6.3.1 The Day–Night Cycle

Although many students articulated the scientific notion of day and night in that their diagrams illustrated the Earth spinning on its axis once every 24 h with the part of the Earth receiving sunlight being in daytime and that receiving no sunlight being in night-time (Fig. 6.2f and g), a substantial number of students (61% of undergraduate and 32% postgraduates) expressed alternative views (see Fig. 6.2a–e). The representations provide insight into reasoning and demonstrate constructs commonly reported by research into pupils' thinking. Some of the ideas have resonance with the mental models of children described by Vosnaidou (1991) such as the Moon and Sun rising and setting as viewed from the Earth's surface (Fig. 6.2). Figures 6.2a–c and e are essentially Earth-centred explanations depicting the Moon and Sun orbiting the Earth in order to create day and night. Figure 6.2b shows the Moon and Sun orbiting a centrally located Earth as an explanation for why we see the Sun and Moon rising and setting. Figure 6.2c also represents the Sun and Moon orbiting the Earth, but reasons that at night the Moon blocks the Sun and as they separate, daytime occurs.

Notions of the Sun being obscured from view during the night-time are common and other studies have shown that explanations involving clouds, hills or moving further into space are evident in pupils across a range of ages (Bailey and Slater 2003). Figures 6.2d and e demonstrate different ideas about the orbits of the Sun and Earth both of which depend on one face of the Earth being illuminated by the Sun in daytime whilst the other face experiences night-time. Trumper (2006) in analysing data drawn from misconceptions studies ranging from Junior High School to University student level, found that the most commonly reported misconception in regard to day and night is that it occurs because the Earth moves around the Sun daily; indeed 51% of future elementary teachers subscribe to this view. Figure 6.2f is essentially correct in that the Earth spins on its axis once every 24 h with the Sun in a fixed position, however, it is incomplete as the Moon is also in a fixed position and is seen as the Earth turns away from the Sun at night-time. On analysing annotated drawings, it was evident that students were employing a range of knowledge. While more than three-quarters of learners indicated that they had knowledge of the Earth spinning, only 33% mentioned the Earth's axis. There was widespread knowledge of the Earth's orbit, but 11 students viewed this as a 24-h orbit of the Sun by the Earth and two students indicated that the Sun orbited the Earth.

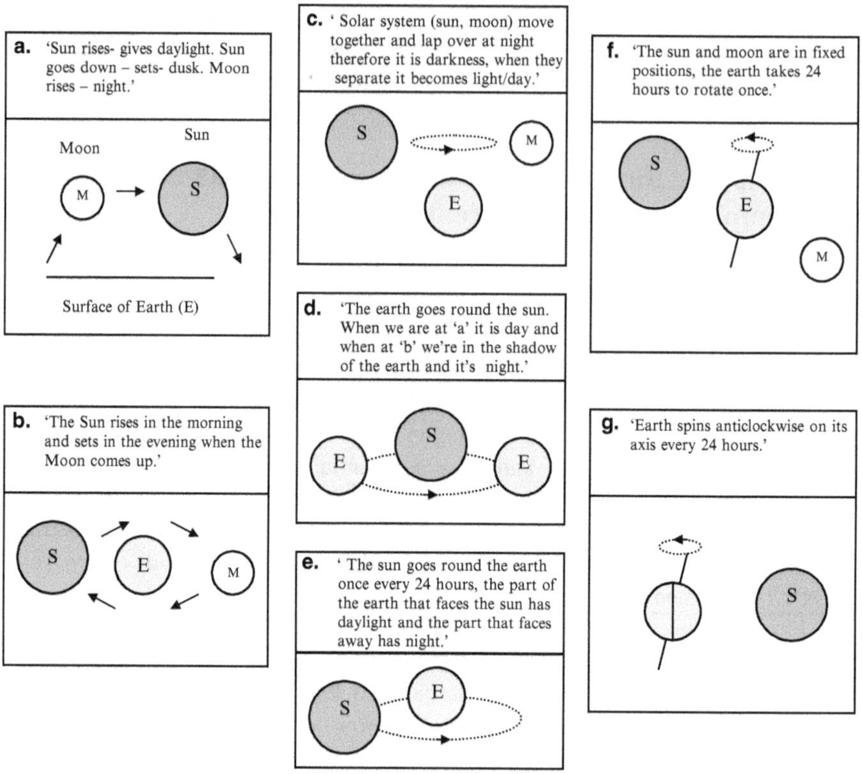

Fig. 6.2 Student explanations of day and night. (**a**) Sun rises – gives daylight. Sun goes down – sets – dusk. Moon rises – night. (**b**) The Sun rises in the morning and sets in the evening when the Moon comes up. (**c**) Solar system (Sun, Moon) move together and lap over at night therefore it is darkness, when they separate it becomes light/day. (**d**) The Earth goes round the Sun. When we are at 'a' it is day and when at 'b' we're in the shadow of the Earth and it's night. (**e**) The Sun goes round the Earth once every 24 h, the part of the Earth that faces the Sun has daylight and the part that faces away has night. (**f**) The Sun and Moon are in fixed positions, the Earth takes 24 h to rotate once. (**g**) Earth spins anticlockwise on its axis every 24 h.

Sharing and discussing the differences between personal annotated diagrams whilst trying out explanations using models to explore causal mechanisms (including the scientific explanation) focused students on the differences between their perceptions and scientific explanation. Stimulus questions were also used, where appropriate, in relation to developing an explanation for time zones, the rotation of the Earth and relative hours of daylight at the equator and the poles. Review of learning is a critical part of raising metacognitive awareness and this was documented through further annotation of original diagrams as well as reflective writing about questions arising, identifying uncertainties and key influences and group discussion of what students saw as significant in their own learning.

Mapping learning revealed that some students who had possessed a scientific view at the outset reported no change in their thinking (they had, however, gained important knowledge about colleagues' reasoning) or had simply assimilated more knowledge, usually about the direction of the Earth's spin, time zones and shadow lengths at various points on the globe but their core constructs had been confirmed:

> I knew the Earth rotated on its axis but I wasn't sure of the direction. By doing the investigation [*of time zones*] it helped clarify and explain day and night clearly in my mind.

Visualisation was paramount in emergent comments:

> Actually seeing it for myself practically and watching the tutor demonstrate it, I could see clearly what was happening and could visualize it in my head.

Many of the students holding clearly alternative views had undergone substantial reorganisation in their thinking:

> I thought that the earth moved round the sun once a day rather than the earth revolving round the sun once a year.

However, like Vosniadou and Brewer (1992), we found that a small contingent merely confirmed their existing ideas or developed them further through incorporating new information, demonstrating how difficult it is to shift entrenched concepts. When faced with anomalous data (see Chapter 4 for further discussion), students sometimes ignore the evidence or incorporate it within existing frameworks as the comments below indicate:

> I was really confident on the reasons for day and night so no explanations were needed [*student had indicated that the Moon blocks out the sun to give night time*]

> I knew that the earth moved around the sun once a day and the sun shone on the earth. I have discovered that whilst half of the earth is in daytime the other half of the earth is in night because 1) the earth is on a tilt therefore the sun can only get to half of it and 2) as the earth goes round one half becomes day and one half becomes night.

6.3.2 The Seasons

The scientific view of seasons (Fig. 6.3c) entails considerably more challenge for the learner. It demands differentiation of the Earth's orbit and spin with respect to the Sun's position as well as knowledge of the Earth's axis in relation to the Sun. The scientific viewpoint relates the amount and duration of sunlight reaching the Earth's surface at different latitudes to the consequences of the planet's tilt.

Only 20% of students articulated key aspects of the scientific explanation of seasons and 27% were unable to formulate a response to the question. Some simply described seasonal change in terms of temperature and day length in various parts of the world. Two broad groups of alternative views were discernible (Fig. 6.3.1 and 6.3.2) and these have been described frequently by other researchers (see Bailey and Slater 2003 and Trumper 2006 for useful reviews). The predominant view involves reasoning based on the relative physical proximity of the Earth to the Sun in summer and winter (*distance model*). In our study a variety of mechanisms were proposed to

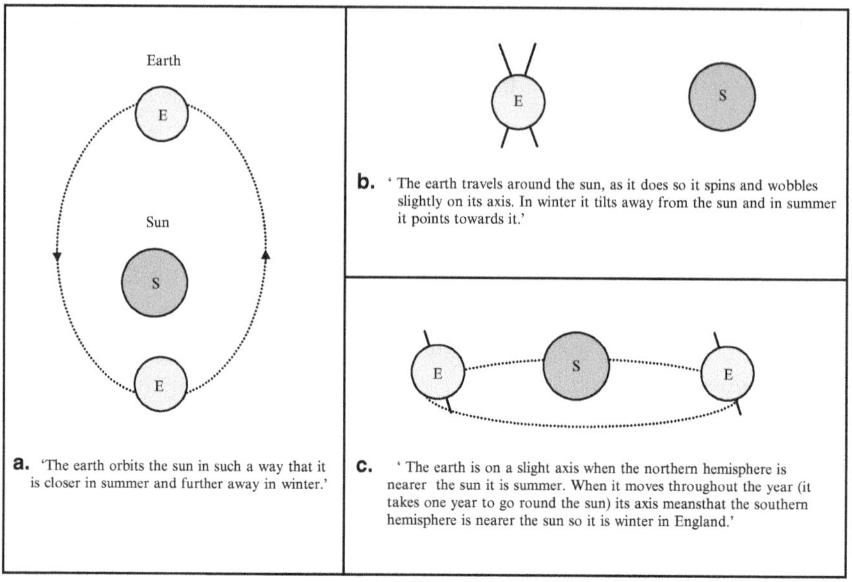

Fig. 6.3 Student explanations of seasons. (**a**) The Earth orbits the Sun in such a way that it is closer in summer and further away in winter.' (**b**) The Earth travels around the Sun, as it does so it spins and wobbles slightly on its axis. In winter it tilts away from the Sun and in summer it points towards it.' (**c**) The Earth is on a slight axis when the northern hemisphere is nearer the Sun it is summer. When it moves throughout the year (it takes 1 year to go round the Sun) its axis means that the southern hemisphere is nearer the Sun so it is winter in England

explain the distance model including elliptical orbits, circular orbits in which the Earth was situated off-centre and spiral orbits and models that entailed the Earth moving backwards and forwards in space. Figure 6.3b represents the second common alternative model in which the Earth's axis appears to oscillate, pointing towards the Sun in summer and away from the Sun in winter (*wobbly Earth model*). The wobbly Earth model is itself underpinned by a distance rationale in that reasoning describes the northern hemisphere being physically nearer the Sun in summer as the axis moves to point towards the Sun. Of the 11 students expressing a 'scientific' view, five used a distance rationale to explain how the tilt of the Earth in relation to the Sun would affect the amount of light or warmth received and only one learner made specific reference to the way in which the direction of the Sun's rays falling on the Earth would affect light intensity. No one attempted to describe why day length changes seasonally in temperate latitudes. So although learners may possess knowledge that the planet's tilt causes the seasons to occur, their reasoning may well be based on the intuitive view of proximity to the Sun with its consequences for temperature and they may fail to apply this model coherently in explaining day length.

As with day and night, it was possible to identify certain categories of how thinking had been affected by learning experiences. First, those students who had a seemingly scientific view of how seasons occur at the outset, had usually extended their insight and incorporated day length into their thinking:

> My previous knowledge was quite good. I learned that the effect of the elliptical orbit of the earth around the sun on night/day length or seasonal sunlight is nil.
>
> I realised why we have longer days in summer and shorter days in winter.

A second category of students, who had expressed limited knowledge at the outset (for example, 'the position of the Sun determines the seasons'), had begun to distinguish the concepts of orbit and spin, and the effect of the Earth's axis in relation to the Sun in determining seasonal change:

> Earth moves round the sun……..axis stays the same! The position of the sun in relation to the earth determines the seasons.

Another discernible category involved students who had a reasonable grasp of orbit and spin but who had developed a distance view of seasons. This group often reported developing insight into the role of the Earth's axis as a sudden occurrence and there were those who had attempted to assimilate new knowledge into their existing frameworks:

> The tilt causes the seasons, I realised that the difference in distance is irrelevant
> [*initial explanation: distance model*]
>
> The complete opposite happens – i.e. the axis does not change its tilt but stays the same.
>
> [*initial explanation: wobbly Earth model*].
>
> I didn't know how we get seasons. I didn't know that the Earth's axis was on a slant. I did know that the Earth went round the sun but until now I didn't know how we get seasons.
>
> [*no initial explanation presented*].

6.3.3 The Phases of the Moon

The cognitive demand of learning is increased considerably with the introduction of another body that is orbiting and spinning in space that the learner must contend with in understanding the phases of the Moon. In common with other studies (for example, Targan 1987; Stahly et al. 1999; Trundle et al. 2002; Mulholland and Ginns 2008), the pre-service teachers found this to be a difficult subject. Only four of the 41 post-graduate students exhibited a scientific view of the occurrence of the phases of the Moon, with 23 expressing alternative views, and five students unable to formulate any explanation. In order to understand how the phases occur according to the scientific explanation, the learner must possess a sophisticated knowledge and understanding of the relative movements of the Earth, Sun and Moon and they will need to appreciate that the phase observed is dependent on the observer's viewpoint within the system. There needs to be specific recognition that despite half the Moon being illuminated by the Sun at all times, we only observe a portion of that (the phase) from the Earth and the phase observed depends on the relative positions of the Earth, Sun and Moon. Furthermore, knowledge of how light travels and is reflected from spherical surfaces is critical in explaining the shape of the shadow observed.

Students demonstrated a strong awareness of the Moon's appearance during its monthly cycle and many were able to supply names for the various phases and indicate its orbit around the Earth. Of the alternative views expressed, most attributed the phases to the 'casting of a shadow' or 'blocking of light' reaching the Moon by the Earth, or less frequently by Sun or other planets. Only five students indicated explicitly that the Moon reflected sunlight. Again, learning outcomes varied, for example some students developed knowledge of the orbits of the Sun, Moon and Earth, but this did not necessarily result in a movement from their initial conception of the phases being caused by the casting of a shadow:

> The moon has its own orbit and spin and that's why the light gets blocked from it.

The realisation that the *relative positions* of the Sun, Earth and Moon precipitate the visible phases was a critical feature of learning for many:

> Viewing the moon and sun from the centre point gave a much clearer understanding – you only see a bit of the moon that's illuminated by the sun!

One question that arose frequently in discussion concerned knowledge students had pertaining to the notion that it is possible to see only one side of the Moon from the Earth. There was ambiguity concerning this idea and the group was keen to develop explanation for it. They were supplied with an explanatory diagram showing the Moon's orbit and spin in relation to other bodies, together with information that the rate of spin and orbit of the Moon are of the same duration. Using this information they proceeded to explore the notion and the following transcript details the thinking of a group of three students (F, M and G) attempting to model the relative positions and movements of the respective bodies. One learner (F) has decided that the time it takes the Moon to spin on its axis is irrelevant data and is currently spinning the Moon on its axis at right angles to the Earth as it orbits:

> As long as one pole [*of the Moon*] faces the earth you'll always have a dark side ... the angle has to be at right angles to the earth. (F)
>
> So that's never going to get any light from the sun? (G)
> No, it depends on where on earth you see it from. (F)
> No the point at which it's facing us will never get any sunshine, that's why its dark. (M)
> It does turn towards the sun... that's the point where you get a lunar eclipse and the dark side comes round and we don't see it. (G)
> You have to see that this will never face that... it orbits the same as it spins . . . It's going within two different dimensions within itself.(F)
> So we get our phases from the orbit not the spin? (M)

The group goes on to conclude that the spin of the Moon is irrelevant; as long as it spins with its axis at right angles to the Earth, there will always be a dark side:

> We did not really think about the statement that the rate of spin of the moon was the same as its orbit and when we came to model it, we forgot about it. This meant we came up with our own logical explanation – the moon's axis always presents a pole to the earth – this fitted and we were happy with it until we realised it was wrong.

In focusing on the spin and orbit of the Moon, while substituting a person in the position of the Earth and making systematic observations at various points in the Moon's orbit using a cross as a reference point on the Moon's surface, students began to recognise the importance of perspective within the system in explaining the phenomenon:

> The demonstration explained exactly a concept I found impossible to understand. I didn't realise this existed [the *spin and orbit of the Moon*], and discovered that you need to keep both in mind together to explain why you see one side of the Moon from the Earth. Using x marks the spot on the ball [*Moon*] proved this concept and I could see why it happened. I needed to be the Earth to see it properly.

The above extract raises several important issues. First, there is the difficulty of holding notions of both orbit and spin simultaneously. This is an area where communicating thinking from one person to another can be problematic. Movements of the bodies are likely to be viewed from slightly different positions and, clearly, communication and interpretation can vary from person to person. This is evident in the group conversation between students F, G and M above. Second, learners experience difficulty in interpreting information expressed in two-dimensional (2-D) diagrams and written form into a three-dimensional (3-D) working model. This was apparent with respect to the question as to why we see only one side of the Moon from the Earth. Despite being provided both verbally, and in diagrammatic form with relevant information about the rate of spin of the Moon on its axis being the same as the time taken to orbit the Earth once, the relatively slow rate of spin remained a conceptual difficulty.

6.4 Insights Identified Through Adopting a Metacognitive Approach to Learning

6.4.1 The Nature of Cognitive Development Within the Subject Domain the Earth and Beyond

The example of student learning illustrated above is testament to the challenges posed by this area of physical science and it raises a central question concerning the process of conceptual acquisition and the extent to which key features of the learning process can be identified to help teachers in constructing and interpreting meaning. Simply subjecting teachers to creative versions of the same story is unlikely to resolve the problem; repetitive exposure to scientific explanation is no guarantee of the emergence of understanding. In a review of astronomy education research, Bailey and Slater (2003) conclude that the underlying cause of student difficulties in astronomy is an area of research needing further exploration. They contend that memorisation is insufficient as a means of 'knowing' or 'understanding' and place emphasis on student-centred approaches to instruction including the need to understand more about student variables and their impact on cognition.

The mechanisms involved in the development of understanding require identification and attempts at this in research to date are inconclusive. Rationalising often-conflicting evidence *for the individual* requires a synthesis of each in order to make sense of a scientific explanation that does not resonate with personal experience of the world. This is a particular problem in understanding basic astronomy in several aspects including:

- the apparent movement of the Sun across the sky
- the imperceptible spin of the Earth
- geocentric view of the Earth's place in the solar system
- the commonsense association between heat source and proximity in explaining warming
- the relationship between the blocking of light and shadow form
- the difficulty inherent in mentally holding and manipulating the orbit and spin of bodies in relation to each other
- the problem of scale and size of even the Earth–Moon–Sun system

A metacognitive approach to pedagogy can offer insight into the mechanisms in which new information is integrated into an existing alternative view and how scientific models of explanation often require the radical reorganisation of thinking from initial ideas. Baxter's (1989) work with children aged 9 to 16 years showed a longer-term progression of ideas from early naïve ideas through notions of astral bodies moving up, down or across, and later embracing the notion of orbital motion. The geocentric position often predominates and a heliocentric explanation requires an abstraction that, prior to the Copernican revolution, confounded science for centuries (Kuhn 1957). It is entirely understandable that making sense of what at first appears, to the scientifically initiated, a straightforward mechanism of explanation through the Earth's spinning on its own axis, is often interpreted by learners through alternative perspectives. A close examination of categories of explanation reveals arrange of seemingly coherent ideas that initially explain observed events. The extent to which these influence a learner's readiness to accommodate an equally coherent alternative perspective seems to be significant and dependent on the way in which thinking is challenged and explored as well as the way in which an alternative perspective is generated or presented and its coherence examined.

There are implications concerning the transfer of ideas in promoting alternative interpretation and the extent to which culturally received wisdom is incorporated. It is influenced by a number of factors including the effectiveness of the constructs developed to bridge scientifically accepted ideas with the individual's underlying conceptual structures, and the coherence of the idea itself. In the case of day and night this could derive from the fact that such a mechanism requires the consideration of only one concept, that of spin and that relative movement with respect to position in space is not a complicating factor. Other observed events do, however, require conceptualising the Earth as a sphere in conjunction with its spin and also its orbit. An understanding of the Sun's apparent movement across the sky demands a synthesis of shape and magnitude (a giant sphere) with movement (spin). Further consideration of periodic change in daylight hours throughout the year in the temperate zones is conceptually much more complex and involves incorporating three ideas simultaneously, namely spin, orbit and tilt, in developing an understanding of a scientific explanation. An account of seasonal periodic change demands the conceptualisation of position and movement in space holding the axis tilt of the Earth constant relative to its orbital plane around the Sun. This compounds the issue significantly and students' initial ideas confirmed that an understanding of the scientific model of seasonal change was considerably more difficult to comprehend and articulate. That is not to say that there is no possibility of developing existing ideas that are in conflict with scientific explanations.

The evidence from the student's accounts of their own learning in the Parker and Heywood (1998) study indicates that movement in cognition did occur. A central question concerns the processes that underpin learning in such circumstances and those elements that contribute positively to accessing conceptual areas. According to Sharp (1996), conceptual change of this order requires a radical shift in thinking from an intuitive egocentric view towards a scientific, remote objective interpretation. It is implied that facilitating the construction of meaning in problems of this nature is dependent on identifying 'hierarchical enabling concepts' in supporting learning. Hierarchical enabling concepts are a similar notion to bridging analogies investigated by Browne (1994a, b) in which he explored facilitating conceptual bridges through presenting bridges using analogies that progressively build on existing experience towards more abstract notions. Such axiomatic reasoning is an attempt to make the abstract more tangible by relating the concept broadly to already established or at least believable ideas. The extent to which this process is transferable to increasingly abstract and difficult ideas as encountered in developing an understanding of basic astronomical events is questionable. It is a pedagogical issue of gradation in explanation synthesised with gradation in concept acquisition.

6.4.2 Using Key Features of Learning to Stimulate the Development of Subject and Pedagogical Knowledge

Insight generated from the student perspective yields valuable information for practitioners on key ideas and influences in the learning of the subject content. Our findings point towards the importance of identifying a structured framework which outlines *key features* in promoting understanding and making these explicit to teachers during their own learning. Key features constitute important pedagogical content knowledge for both tutor and intending teacher; they concern identifying the processes and conceptual frameworks that underpin qualitative constructs for causal explanations and constitute important pedagogic content knowledge. The significance of this is outlined by Shulman (1987) as involving much more than the explication of subject knowledge. This is where a metacognitive awareness becomes an invaluable tool in both learning and teaching. Several key features of the learning process identified from the student perspective included not only the need to be confronted with the key scientific ideas, but also the importance of being made explicitly aware of those factors which promote access to such ideas and elements which are restrictive in the development of understanding. The significant key features instrumental in the construction of causal explanations for the astronomical events investigated represent a synthesis of subject. They included:

6.4.2.1 Scale and Spatial Awareness

> Scale is important – how big everything is (the sun compared to the earth) and how far away they are.

First, there is the generic problem of spatial awareness in relating to position in space of the observer and the observed objects. Students constantly expressed a need to visualise and clarify how bodies move in relation to each other and what the outcomes are in terms of observable events:

> [T]he physical demonstration of standing around the moon and seeing for myself the shape of the visible light was a revelation!

> I was unaware that (a) the moon made only one spin during its orbit of the earth and (b) that you only saw one side of the moon. Even with this information I couldn't accept it until I saw it in the class demonstration.

Practical modelling was paramount in learners clarifying and articulating ideas and an integral part of this process involved listening to and observing others' explanations. Visualisation through modelling facilitates access to the scientific model in that the learner is able to see for themselves scale, spatial relationships and relative movements and is provided with a medium through which they can consider a range of solutions and the degree of coherence with the currently accepted explanation. Important pedagogical knowledge can be derived from the experience such as the potential to stimulate misconceptions in terms of scale and dimension when using a globe to represent the Earth and a torch to represent the Sun as commonly occurs in elementary classrooms.

6.4.2.2 Two- and Three-Dimensional Reasoning

> The 2D picture confused me. I initially thought I understood this picture yet with a closer look I found it very difficult to visualise.

Translation of 2-D diagrams into meaningful 3-D models emerged from reflections as a key obstacle in the interpretation of seasons and the phases of the Moon. Learners often experienced difficulty in constructing 3-D models from 2-D description and needed considerable support in clarifying relative positions and movements in space. In translating 2-D information accurately a learner must synthesise information regarding relative positions and sizes of the bodies involved, their movements (orbit and spin), their orientations and their rate of movement *in relation to each other*. Furthermore, the position in space (position of the observer) from which the dynamic system is being viewed is critical in interpreting the information.

Culturally received science from textbooks illustrates the point, since in Western art most representation is orientated from left- to right-side elevation which is a particularly distorting view of the frame, freezing the movement of spin and orbit. Barnett et al. (2001) also found this to be significant:

> This difficulty in developing an understanding of astronomical concepts arises, in part, because the science of astronomy requires students to develop an understanding of the complex relationships and dynamics between objects in 3-D space. In addition, students are required to examine, mentally or otherwise, objects and events from multiple perspectives (Parker and Heywood 1998). However, despite the 3-D nature of astronomy, most resources available to students are in the form of 2-D charts and images within textbooks that attempt to emulate astronomical phenomena from different 3-D perspectives. (Barnett et al. 2001:8)

Reflection on learning allows students to recognise their own mistakes in interpretation, thereby deriving important pedagogical insights for future practice:

> It was interesting that although the diagram showed clearly that the axis remained constant *[the Earth's tilt in seasonal change*] when we modelled this we didn't keep it the same and we didn't realise that's what we'd done! The teacher needs to constantly check how pupils understand diagrams, if the tutor hadn't intervened and made us really look at what the picture meant we'd have just continued.

> 2D representations of 3D objects at different positions makes this subject difficult. You need to be clear on the positions in space and how bodies are moving in relation to each other.

In interpreting the phases, the learner needs to position himself/herself as an observer looking at the Sun and Moon and to recognise that what is seen is dependent upon the relative positions of the observer (in this case from Earth), the reflective object (the Moon) and the light source (the Sun). In order to explain the shape of the Moon phases scientifically, the learner needs to have experienced how shadows are formed when light is incident on a sphere when observed from different positions and angles; there is only one set of circumstances when a half moon can be observed. We would suggest that the difficulties experienced in formulating an explanation of this observed effect is explicitly illustrated in attempting to discern whether the Moon will look the same in Australia! A key recognition made by students in this regard concerned was associated with the observer's perspective. In placing the observer in the position of the Earth with the Moon represented by a small white ball (in order to show shadow lines clearly), it became apparent that the shape of the phase was determined by the amount of illuminated Moon surface that could be seen from the Earth:

> Viewing the moon and sun from the centre point gave a much clearer understanding.

> To demonstrate the moon phases you need 3 components: light (sun), a ball (moon) and a person (Earth).

Given the rapid advances in digital technology witnessed in recent years, the teacher now has at their disposal extensive resources with which to model in 3-D (see for example, Roth 1996; Barnett et al. 2001; Barab et al. 2002; Keating et al. 2002; Yu 2005; Mikropoulos 2006). Modelling provides a medium and the cognitive tools with which to represent, enact and translate ideas in science where it is not possible to access subject matter first-hand. Barab et al.'s Virtual Solar System, for instance, is used to support students in visualising size, spatial relationships and relative motion of bodies as learners think critically about basic astronomical phenomena as part of specific learning activities that place the learner as meaning maker at the centre of the learning process:

> While constructing models, a "conversation" unfolds through which interactions occur among students, among students and the teacher, and among students and their models and the materials (inscriptions) of their work, as students attempt to create meaning through and from their constructions (Roth 1996) (Barab et al. 2002:77)

Students become engaged in an iterative process between evolving explanation and the development and testing of models. This requires clarity in regard to the observer's perspective; their interpretation of their vantage point in viewing the system is critical and instruction must address this explicitly. The metacognitive approach towards learning succeeded in highlighting this critical feature in the concrete modelling context but such insight might also inform digital representation such that students are able to interpret systems more accurately. In this process mental images play a central role. Mathewson (1999) provides a detailed discussion of visual-spatial reasoning and highlights its importance in developing higher order thinking. Visual-spatial thinking combines vision (using the eyes to identify and think about an object) and imagery (the formation, maintenance and transformation of images in the mind in the absence of the object). Such mental images are to the learner coherent encodings of experience and constitute the most integrative of mental processes, characterising the expert user of knowledge. Mathewson presents a convincing case for a stronger educational focus on developing visual-spatial thinking in learning, particularly at the time when language flourishes. This is supported by Gilbert (2005) who claims that the process of simplification and representation within the scope of human senses with the aid of models is of great importance in understanding science and that models are vital in the visualisation (visual imagery) of entities, relationships, causes and effects. He argues that as the role of models and modelling is a fairly recent event in science education, teacher training needs to sustain a more model-based curriculum such that trainees' science content knowledge includes a comprehensive understanding of 'curriculum models' including range, appropriateness and the nature of their inherent cause–effect relationships. It is important that teachers understand the scope and limitations of models so that they might judge their educational efficacy in relation to context. The exercise of metacognitive awareness and self-regulation in the act of making pedagogical decisions regarding models and their application in the act of teaching and learning is essentially a metacognitive process.

6.4.2.3 Spin and Orbit

> I got mixed up in the earth turning itself and the earth moving round the sun in one year.

Many learners failed to differentiate between the notions of spin and orbit in both conversation and writing. This is not to say that they did not do so mentally, but the lack of precision in communication is a possible barrier to be reconciled in teaching. There was a high degree of uncertainty in relation to the language as to what constituted 'orbit' and how this differed from 'spin'. Spin and orbit cannot be represented adequately for learners in the 2-D format but even where 3-D models of representation are used, most found difficulty in applying the notions to other situations (such as in explaining why only one face of the Moon is seen from Earth). Oscillating from one medium to the other can prove a useful strategy in consolidating ideas, but in teaching this needs cautionary treatment since it is easy to confuse. The problem is not necessarily resolved through breaking the whole into parts, no matter how carefully structured. An intuitive sense of spin assumes a relatively *fast* movement and this forms an obstacle in translating, for example, the relatively slow spin of the Moon. It is not unproblematic, therefore, to distinguish spin and orbit in translating 2-D form into 3-D movement; however, the fundamental difficulty lies in the need to *simultaneously* consider each *in relation* to the other when developing a coherent understanding of basic astronomical events such as the Sun, Earth, and Moon system. The recognition of this problem is conspicuous by its absence in pedagogical research literature and, consequently, the metacognitive approach generated unique insight into an aspect of learning of significance in real teaching situations. The ability of the learner to synthesise orbit and spin could well be a result of an assumed culturally accepted explanation to which 'everyone in the know' subscribes and makes sense of in an unproblematic way. Clearly this is not the case. Vosniadou suggests that the distance model of the seasons is actively promoted in some texts with such phrases as 'when the earth tilts towards the Sun it is summer and when it is further away from the Sun we have winter' (Vosnaidou 1991: 232). No such reasoning could account for a change in daylight length. What is needed is a pedagogic structure that facilitates a visualization of the mechanism that accounts for all the observed phenomena and experiences associated with seasonal change, and this includes the changing duration of daylight.

Through developing metacognitive awareness of the problem of failing to distinguish orbit and spin and to hold the notions simultaneously in translating ideas and diagrammatic explanations, students are better able to exert control over future learning through recognising the need to be precise in determining spin and orbit rates of bodies relative to each other:

> I didn't realise this existed [*slow rate of spin of the Moon*] and wouldn't have thought about it ordinarily, you need to relate it to phase of orbit to explain it.

6.4.2.4 The Earth's Tilt

> I don't think that I ever thought that the earth is on a tilt even if I have been told.

It may seem less than obvious that both spin and orbit need to be considered together in understanding change in daylight length throughout the year. To understand this it is necessary to 'visualise' the tilt of the Earth's axis at different points in the Earth's orbit around the Sun. This is best exemplified at four strategic points, the summer and winter solstice and the autumn and spring equinoxes. In modelling a 3-D demonstration, a key teaching point to emphasise is the need to keep the tilt of the Earth's axis constant to the orbital plane throughout the entire orbit; something which many learners find difficulty in doing. The key that unlocks the puzzle concerns the duration of spin in shadow and light:

> The key is in the tilt – it stays the same-but it's hard to see that in a diagram.

> I had difficulty understanding seasons, it became clear when I sat behind the beam and we turned the globe around to represent orbit. This way I could see where the shadow was, with more time in daylight in summer and less in winter because of the earth's tilt. Seeing the beam shining on the earth helped me to understand its spread and intensity in certain places.

Thus in winter, in the northern hemisphere, when the tilt of the axis is pointing away from the Sun, the duration of turn in shadow is greater than the duration of turn in the light. This equates with short daylight length in winter. The reverse is true in summer. Of course in the southern hemisphere the same orbital position results in a greater duration of turn in the light. The effect is exaggerated the further from the equator. The visual effect of seeing how much time any one location spends in daylight and darkness as a result of the Earth's tilt at key points in the Earth's orbit of the Sun constitutes an essential aspect of building understanding of why day length changes seasonally:

> Once we had a white ball for the earth I could see the shadow and light changes and I could understand. I had misunderstood the tilt, I thought it changed in orbit but it doesn't.

6.4.2.5 Light Shining on a Sphere

> Most important was actually seeing the shadow line made by the projected light.

Another matter of significance identified from the student perspective concerns the nature of shadow formation on a sphere. There needs to be explicit recognition of how light shining on the spherical globe model creates a vertical shadow line which does not follow the tilt line of the axis (as illustrated in Fig. 6.1). Students identified the experience of seeing this shadow line as critically important in demarcating day length at various positions in the Earth's orbit of the Sun:

> [*A critical aspect of learning was*].. the shadow proportions shown by the globe facing light and changing positions of light/globe and freezing positions.

We have found that if this is not explicitly highlighted in teaching then understanding is more likely to remain at best partial. Not only is it important that students observe the vertical shadow line produced as light strikes a sphere, but the recognition that this vertical line appears different when viewed from various positions and angles (as evidenced by the phases on the Moon) is critical knowledge in underpinning the development of causal explanation. Again, what appears to be a fundamental prerequisite for understanding seasons from the learner's perspective is undervalued in current literature. Herein lies an important point, that in developing metacognitive awareness of learning, students are able to engage tutors in further professional insight into pedagogy. This becomes a two-way process of professional endeavour.

6.4.2.6 Language and Communication

> Understanding the language is very important in understanding the concepts

There is a considerable language issue in developing knowledge and understanding in this domain and students quickly became aware of the potential for miscommunication and the consequent lack of accuracy in instruction and assessment of learning. Students were able to make quite subtle observations on their own use of language:

> The language we use is crucial. For example, the s.hemisphere is closer to the sun, not the sun is closer to the s. hemisphere in understanding how the tilt of the earth causes seasons.

The evident difficulty in describing movement in space, the use of everyday language, the interchangeability of terms and lack of precision in regard to 'spin' and 'orbit' were identified as particular barriers in learning. The scientific use of orbit and spin was not recognised as significant early on in instruction and students' initial ideas failed to employ the word 'spin' systematically and few mentioned orbit in any consequential way. Atwood and Atwood (1995) in exploring pre-service elementary teachers' conceptions of the causes night and day, noted that participants often used the word 'rotate' inappropriately in their explanations, for example 'Earth rotates around the Sun'. However, with use of physical models, the same participants often clearly indicated the motion scientifically defined as 'revolving' or 'orbiting'. The students in our study came to recognise this as an acute problem, particularly in developing understanding of the phases of the Moon where the words spin and orbit with respect to the cyclical movements of the Moon hold particular difficulties. This culminated in discussion of the existence of the dark side of the Moon where a causal explanation as to why you see only one face of the Moon from Earth, even though the Moon is spinning on its own axis, presents considerable challenge for most learners. It is summarised in the following statement:

> The reason you see only one face of the moon from earth is because the moon's phase of spin is the same as its phase of orbit.

This challenge requires addressing the language issue discussed, the interpretation of spin and orbit, and the ensuing difficulties encountered in three-dimensional movement and position in space. An understanding of this requires a review of the interpretation of the word spin because in this example the spin rate is very slow compared with most people's personal experience of spinning objects. Spin is often associated with rapid rotation; a spin that takes approximately 28 days to complete is therefore difficult to envisage. The students in our study experienced considerable difficulty in accessing an interpretation of this and although the above statement is a 'correct scientific content knowledge' description as to a causal explanation of what is observed, it fails to provide meaningful explanation. The inadequacy of such an explanation exemplifies our very point and is the essence of our argument, that it is not the *knowing* of the explanation but the *understanding* of the factors contributing to the development of understanding that is the key issue for teachers and learners.

6.5 Discussion

This chapter has focused on the potential of metacognition in generating subject-related pedagogical knowledge of simple astronomical events as pre-service teachers strive to develop causal explanations. It illustrates how metacognition affords the opportunity for both students and tutors to gain unique insight into the learning of subject in this area with important emergent implications for instructional practice.

A fundamental principle of science teacher education should be concerned with identifying and making explicit the underlying conceptual frameworks that the learner is likely to have difficulty becoming encultured into the scientific interpretation of events. A metacognitive approach such as that described above is most effectively achieved through teachers auditing their own learning while engaging with the concepts themselves. This is essentially different from simply addressing the issue of knowledge acquisition for teachers with alternative conceptual frameworks; it requires the development of metacognitive capabilities in generating pedagogical insight into subject. In terms of Flavell's (1987) conception of *metacognitive awareness* or Schraw's (1998) categorisation of *knowledge of cognition*, the exemplification provided here represents a body of pedagogical knowledge, the generation of which is contingent on students' abilities to identify and reflect on or 'think about' their own learning. This demands engaging at a higher cognitive level such that they begin to raise implications for themselves as learners and teachers of science. Through this process, pre-service teachers are made more aware of the pedagogical implications for practice, and in the next chapter we illustrate how this knowledge impacts more explicitly on approaches towards planning for the teaching of subject.

Through the process of 'problematising' subject matter, learners become aware of the cognitive landscape of learning in particular subject domains thereby identifying cognitive difficulties presented by the subject matter in question (Heywood 2007). Some of this knowledge about the learning of basic astronomy, such as the challenge of synthesising orbit and spin, the importance of prior consideration of how light shines on a sphere and the importance of the observer's perspective, has been underestimated in educational literature and, yet from the student perspective, they constitute critical elements of learning. Such knowledge emerges from both an awareness of themselves as learners and knowledge of the factors influencing learning (declarative knowledge). As a direct consequence of this experience, students identified the need to consider carefully the use and interpretation of associated language and 2-D to 3-D representation (procedural knowledge). In terms of teacher education, this knowledge is likely to have an impact in the actions of planning and teaching of subject in the classroom (conditional knowledge). In regard to self-management or regulation of learning, not only did students develop knowledge of their own language use and interpretation of the movements and spatial and positional importance of bodies within the systems explored, they also identified issues of communication in terms of diagrammatic representation and how language serves to focus attention on key aspects of problem solving as they sought to model these phenomena. Wenden (1999) discusses the relationship between metacognitive knowledge and beliefs and self-regulation (or self-direction) in language learning where learners' acquired knowledge about learning has been shown to influence task analysis and monitoring in self-regulation. She contends that engagement in pre-task planning involves learners in calling upon their metacognitive knowledge or beliefs about a particular task to identify the nature of the problem it poses and to consider similarities to current experience in determining how to proceed. In teacher education, where the task ultimately translates into planning for the learning of children, metacognitive knowledge and beliefs have potential to inform planning in terms of considering the nature of subject matter and related pedagogy that the teacher considers best suited to supporting learning.

Without insight into the intricacies of the learning process, subtle influences that impact markedly on the capacity for learners to make the necessary conceptual links are likely to be lost. For teacher education it is not simply a case of understanding the conceptual domain of the Earth's spin causing day and night, or the orbit and tilt causing the seasons, it is also necessary to recognise the nuances of structuring learning to make the conceptual frameworks accessible for learners. Recognition of that structure through the individual learning experience is a two-way, dynamic process that involves both tutor and student in the generation of pedagogy. Georghiades (2001) views metacognition not as something 'taught' to the learner in an 'outside-in' process, but rather it is a skill that can be helped to develop in an 'inside-out' manner. In science teacher education, developing the capacity for self-reflection on the learning process in specific subject domains affords the opportunity to develop the skill of being able to identify similarities and differences between subjects themselves. The holding of two concepts, for instance, (such as spin and orbit) in a relationship and the cognitive challenge this presents is not exclusive to astronomy. The same problem is faced in other physical science domains such as synthesising current and energy transfer in understanding simple electrical circuits (Chapter 3) or weight and size in floating and sinking (Chapter 2). This pedagogic insight transcends mere subject knowledge and is dependent on the development of metacognitive skills. The implications of this for teacher education curricula will be explored further in the concluding chapter of this book.

Chapter 7
The Subject Matter Learning Audit and the Generation of Pedagogical Content Knowledge

The previous chapters of this book have illustrated how developing metacognitive awareness of learning in the physical sciences holds potential for generating insight into not only the nature of science subject matter, but also the learning of subject with distinct advantages for self-regulation of personal learning and emergent implications for teaching. This chapter explores the generation of pedagogical insight through the application of a Subject Matter Learning Audit (SMLA) approach and presents empirical case study data showing how pre-service teachers, at an early stage in their professional training, employ knowledge of their own learning of subject in the formulation of teaching. We argue that personal learning experience is an undervalued avenue for the development of science pedagogical knowledge and conclude by discussing some important implications for the education of primary school teachers.

7.1 Teacher Knowledge

We all possess opinions as to what constitutes the 'expert' teacher, the teacher who inspires pupils and who is able to use pedagogic knowledge almost effortlessly to secure effective educational environments in which learners are nurtured intellectually. However, as Turner-Bisset (2001) points out, defining precisely what 'expert' teaching means is problematic as there is a lack of consensus in the literature as to not only the terminology used to describe it, but also how such expert teaching is to be defined and assessed using appropriate criteria. The English schools inspectorate (Ofsted) grade teaching subjectively as 'outstanding', 'good', 'satisfactory' or 'inadequate' in their assessment of teacher performance. In a report of teaching in a school recently given the highest grade of performance they state:

> Teaching is outstanding throughout the school. It is typified by a number of very strong features, including: high expectations of what pupils should achieve; very good levels of support from teaching assistants; well-planned lessons that make clear what pupils are to learn; and activities that are interesting and matched to need. Information and communication technology is used very well to support teaching and make learning fun. Pupils participate

D. Heywood and J. Parker, *The Pedagogy of Physical Science*, Contemporary Trends and Issues in Science Education, vol. 38, DOI 10.1007/978-1-4020-5271-2_7,

> well and particularly enjoy the opportunities to discuss with their talking partners what they are learning; for example, clock times. A key feature of lessons is the way in which teachers and their assistants question and encourage pupils to think and give that little bit more. As a result of these strengths, lessons are pacy, develop speaking and listening skills well, and ensure that pupils make excellent progress in their learning. (Ofsted 2008)

At the heart of these comments lies the ability of the teacher to understand the educational needs of the children they work with and be able to plan the next appropriate step to provide lessons that engage them in learning. A key feature of lessons resides in the effectiveness of teacher questioning in terms of encouraging pupils to think further. In other words, effective teachers possess a subtle knowledge of subject, learners and the learning process and have the ability to plan, organise and control the classroom accordingly.

What knowledge bases contribute to such expert teaching and what are the implications of this for how we train our teachers and support them in their continuing professional development? To pin down what is, in essence, tacit knowledge, constructed as a result of individual, unique experiences of both personal learning and classroom practice, presents a significant challenge. A variety of approaches have arisen to categorise and describe its various manifestations (see for example, Ball and McDairmid 1996; Calderhead 1996; Munby et al. 2001; Turner-Bisset 2001; Abell 2007). Abell (2007) presents an historical perspective of teacher knowledge in which she explores some of the meanings for the term subscribed to in educational research over the past 50 years. Drawing on analyses by Fenstermacher (1994) and Reynolds (1989), she discusses its evolution within science education beginning from a position of teacher knowledge as a static component (such as a qualification or competency) that is compared with practice or learning outcomes. Research of the 1980s and 1990s illustrates a paradigm shift from the teacher as 'knower' to 'teachers' practice knowledge' (Connelly and Clandinin 1990; Cochran-Smith and Lytle 1999; Schön 1987). In the former the teacher's knowledge base constitutes the emphasis of research, whereas the latter is derived from teachers participating in the act of teaching.

Teaching is a profession informed by a range of knowledge bases. In terms of subject these include *substantive knowledge* (knowledge of the facts and concepts of a discipline), *syntactic knowledge* (knowledge of skills and processes of the discipline) and *beliefs* (including attitudes and values) about the teaching of that discipline (Schwab 1964, 1978). Turner-Bisset (2001) identifies the kinds of knowledge historically required for teaching as:

- academic subject knowledge
- craft knowledge
- knowledge of a particular moral code
- curriculum knowledge (what is to be taught)
- knowledge of educational theory informed by the four foundation disciplines
- knowledge of child development (Turner-Bisset 2001:10)

She reflects that the relative importance of these kinds of knowledge waxes and wanes over time due to dominant politics, fashion and educational ideologies and argues that recent times have witnessed a move away from academic knowledge towards a focus on skills and processes of teaching. A recent report on English school's performance in relation to primary science education found that schools placed emphasis on the role of science enquiry in children's learning and, in particular, the opportunity to pose questions, design, and carry out investigations for themselves (Ofsted 2007). The report considered that in planning science activities, teachers did not take sufficient account of what children had already learned and did not give them clear advice on how to improve their work further. It recommended that continuing professional development needs to focus particularly on science knowledge, understanding and progression in learning. In a sense this suggests a need to redress the balance of teacher knowledge bases in regard to science subject content in interpreting the curriculum in order to enable science teachers to plan more effectively for children's learning.

7.1.1 Pedagogic Content Knowledge

The conceptualisation of teacher knowledge developed by Shulman and his colleagues has proved to be highly influential in science education research. The importance of this work resides in the distinction it makes between general pedagogic knowledge and pedagogic knowledge related specifically to the teaching and learning of topics (Shulman 1986, 1987; Grossman 1990; Wilson et al. 1987). Shulman defined knowledge relating to the translation and representation of subject in teaching as pedagogic content knowledge (PCK). This represents the knowledge developed by teachers to make the subject matter of particular topics accessible to learners and to support the development of their learning through providing for progression. Such knowledge is derived from the act of teaching specific subject matter and informed by teachers' own knowledge of subject (subject matter knowledge), as well as a range of general pedagogic knowledge that transcends subject such as knowledge of pupils, contexts, curricula and general educational purposes. Unsurprisingly, given the inherent difficulty in the teaching and learning of some scientific concepts, the construct of pedagogic content knowledge is of particular interest to science education and science teacher education research where it concerns issues associated with the interpretation and representation of the subject in teaching. PCK encompasses a host of issues such as knowledge of learners' conceptions, difficulties encountered in the learning of subject matter, and knowledge of how to help learners construct understanding through the use of effective instructional approaches (such as scaffolding), and representations (for example using metaphors and analogies).

Shulman's work underpins a range of different conceptualisations and classifications (see for example Marks 1990; Fernandez-Balboa and Stiehl 1995; Adams and Krockover 1997; van Driel et al. 1998; Gess-Newsome 1999; Magnusson et al. 1999; Barnett and Hodson 2001; Appleton 2006). Although a full discussion of these perspectives is beyond the scope of this book, any attempts to research PCK invariably lead towards the isolation of elements that can be accounted for and descriptions of PCK differ in terms of the breadth of such definition. Turner-Bisset (1999, 2001), for instance, views PCK as an all-embracing knowledge base composed of interacting sets of knowledge bases. This encompasses substantive and syntactic knowledge of subject, knowledge of curriculum and beliefs about subject. It also involves knowledge relating to teaching generally, including general pedagogical knowledge, a repertoire of different teaching models, and knowledge of teaching context. Knowledge related to interpersonal interactions in teaching, cognitive and empirical knowledge of learners, knowledge of self and knowledge of educational goals are integral to Turner-Bisset's view. While some authors specify components of PCK, others take a more holistic perspective. Magnusson et al. (1999) employ a broad view of PCK involving five components including: orientations towards science teaching (including knowledge of teachers' goals for and approaches to science teaching); knowledge of the curriculum (at all levels); knowledge of assessment (timing and strategies); instructional strategies (including representations, contexts, activities and methods) and knowledge of learners' science understanding.

Loughran et al. (2002) point to the difficulties of isolating elements of teacher knowledge in research situations as teachers possess a holistic or integrated understanding of their work. Despite conjecture about conceptualisations of PCK in terms of precise definition, and whether it describes a mixture or synthesis of knowledge types, several unifying features of PCK have emerged. For example, van Driel et al. (1998) identify two key elements of PCK studies as being concerned with learner conceptions of particular science subject matter and the difficulties they encounter in the learning of it, as well as knowledge relating to effective representation of subject for learners. However, whatever the problems inherent in articulating PCK, describing it does not show the process by which it is developed (Munby et al. 2001; Mulholland and Wallace 2005). PCK generation begins with teachers as learners of science and evolves as they gain professional experience of teaching and schools. Little is known of PCK development throughout a teacher's lifetime as evolved knowledge is likely to reside within individuals and become lost as teachers retire from the profession. Likewise we know little of the emergence of PCK in the early stages of training and its subsequent evolution within the individual context. Bucat draws attention to the 'vast difference between knowing about a topic and knowing about the teaching and learning of that topic':

> Now we have encyclopaedic collections of student misconceptions, but usually no more than bland, general statements about preventative or curative actions. We have an enhanced knowledge of the conditions for effective learning, based upon which a range of student-centred teaching methodologies, such as cooperative learning, have become fashionable – but little guidance as to how teachers might apply these to the teaching of particular chemistry topics such as reaction kinetics or stereochemistry. Educational research has had little impact on science teaching. (Bucat 2004:1)

Whilst acknowledging that in some sense the teaching skills of outstanding teachers are generic in nature, Bucat contends that many of their pedagogic skills are content-specific. He considers that subject matter has been used as a vehicle for pedagogical ideas and theories (such as constructivist approaches, metacognition and co-operative learning) that transcend the subject matter per se.

7.1.2 Teacher Education and the Development of PCK

PCK development is, necessarily, both individual and context-specific, determined by the nature of the topic, the teacher's perceptions of the topic and the learning of it, and their reflective competency in terms of learning from the act of teaching. Within teacher preparation programmes opportunity for science PCK development occurs within disciplinary education, through independent development of personal knowledge of science subject content and in gaining knowledge of learners and learning through observation and the practice of teaching (Grossman 1990; Sanders et al. 1993; Lederman et al. 1994; De Jong et al. 2005). Research has naturally focused on PCK development as a result of teaching specific science topics in school where pre-service teachers can engage directly with learners' conceptions and recognise difficulties pupils encounter in the learning of subject matter. It is here that they are able to explore instructional strategies and build their repertoire of skills in translating science subject matter for learners. However, as De Jong et al. (2005) point out, such research studies are few and not much is known about the process of PCK development among new teachers, and how its development might be facilitated. In their study of pre-service secondary teachers' PCK of using particle models, they found that starting an educational module by focusing on explicating teachers' initial knowledge of secondary school conceptions and learning difficulties, and expanding and analysing relevant chemistry texts, appeared to stimulate teachers' thinking about potentially useful instructional strategies.

PCK research studies are usually located within the context of teaching and frequently concern the development of PCK in the preparation of secondary science teachers. As Poulson (2001) points out, Shulman recognised that his work was focused on the secondary subject specialist teacher and, as such, its applicability to the primary context with its largely non-specialist work force requires careful interpretation. In primary teacher education the opportunities to develop subject-related PCK are necessarily limited within the broad curriculum, as is the opportunity for science teaching during school placements. Also, the timing of placements and institution-based preparation and follow-up sessions are not always conducive to the in-depth critical review needed to stimulate and expand science PCK. As generalists, primary teachers' PCK may be regarded as weak or more dependent on general pedagogic knowledge or professional knowledge relating to the teaching of a range of subjects (Sanders et al. 1993). Despite these limitations, teachers are expected to plan for children's learning in physics domains that contain challenging concepts such as gravitational attraction, seasonal changes in basic astronomy and the behaviour of simple electrical circuits as we have seen in the preceding chapters. In a 10-year study of PCK development in a teacher making the transition from pre-service to experienced teacher, Mulholland and Wallace (2005) found that while science subject knowledge is initially important, general teaching and interactive knowledge soon overshadow it. They conclude that the science subject knowledge base that primary teachers begin teaching with needs to be sufficiently robust to survive the time when the beginning teacher strives to develop general teacher and interactive knowledge bases. If subject knowledge is not robust, PCK may atrophy to a state where only the accumulation of scientific information is attempted in practice. Harlen and Holroyd (1997), in a survey of Scottish teachers' confidence in teaching science and technology in primary schools and the influence of their scientific understanding, reported that most teachers lacking in confidence employed a variety of 'coping' strategies. These included for instance, compensating for doing less of a low-confidence aspect of science teaching by doing more of a higher-confidence aspect (for example, stressing process skills at the expense of concentrating on conceptual development), placing heavy reliance on kits, prescriptive text and pupil work cards, emphasising expository teaching, underplaying questioning and discussion and avoiding all but the simplest of practical work.

The assumption that a strong knowledge base automatically results in more effective practice is not without question (Cochran-Smith and Lytle 1999; Poulson 2001). Although van Driel et al. (1998) consider a thorough and coherent knowledge of subject matter acts as a prerequisite for developing pedagogic knowledge in relation to the teaching of specific subject content, the realisation of this in practice is fraught with difficulty.

7.1.3 Translation and Interpretation: Knowledge into Practice

The process of translating subject knowledge into effective practice involves teachers in interpreting science ideas for themselves, and then attempting to translate these ideas for pupils through adopting appropriate classroom practices. In teaching, it is necessary to both interpret ideas and focus on the nature of explanations. In the former case, the teacher could be considered the principal interpreter of science (Eger 1992a; Heywood 2002), whilst the latter is perceived ostensibly in terms of pedagogy. To consider these elements as mutually exclusive obscures the complex symbiotic relationship that provides insight into the emergence of effective practice in science teaching. It is here that *learners' perspectives* of subject matter and their learning of it can potentially have significant influence in informing practice. Knowledge of subject matter and knowledge of related pedagogy is, therefore, a complex synthesis and the process through which this professional insight is developed constitutes a central concern for teacher education programmes (Summers 1994). Generating such curricular expertise is necessarily concerned with issues such as developing knowledge of:

- science subject matter
- the curriculum and its interpretation
- learners' initial conceptions and their significance in learning
- the nature and progression of ideas and how learners might build concepts

- strategies that are effective in promoting learning (for example, scaffolding, the role of scientific investigations)
- how to represent ideas for learners using appropriate metaphors, analogies and explanations
- specific difficulties learners are likely to encounter in understanding ideas (such as language issues, counterintuitive scientific ideas)
- beliefs about learning and teaching

A range of factors are likely to impact upon evolving pedagogical knowledge, including a teacher's epistemological beliefs. Personal orientation regarding how best to teach science and how learning takes place is known to be a powerful influence in shaping classroom approaches (Appleton and Asoko 1996; Lunn 2002). Smith and Neale (1991) defined several types of orientation towards science that might lead teachers to prioritise certain approaches in practice including a didactic/content mastery view whereby science is viewed as a body of substantive knowledge, and a conceptual change view of science as a process of personal construction and evolution of theories. Such orientations influence how an individual teacher delivers the science curriculum and, importantly, how the teacher influences the building of images of and orientations toward the subject in the minds of the learners they teach.

The following case study of three pre-service teachers' learning about forces and planning for teaching in this domain represents the focus of our current and ongoing research into the development of science PCK in pre-service teachers. It explores the possibility of stimulating PCK development at an early stage in professional education through what we have termed a Subject Matter Learning Audit (SMLA) that is based on the metacognitive approach to learning described in previous chapters of this book.

7.2 The Subject Matter Learning Audit

7.2.1 Rationale

As opportunities for PCK relating to science topics are necessarily restricted in primary teacher education, the role of institution-based taught sessions within training programmes becomes paramount in nurturing embryonic science pedagogy. The Subject Matter Learning Audit (SMLA) is an approach devised to maximise the PCK dimension of teacher education. Its purpose is to provide a structure that enables pre-service teachers to interpret the curriculum critically from the learning perspective such that they begin to consider the cognitive landscape of learning of a particular domain as they anticipate planning for teaching. The process is dependent on the development of metacognitive awareness of personal learning as the pedagogical stimulus to planning for pupils' learning. In focusing attention on the nature of the subject matter to be taught, there is opportunity to recognise potentially challenging aspects, to consider the reasons why such difficulties might arise and to gain a sense of abstract and counterintuitive ideas involved in constructing scientific explanation. The nature and roots of misconceptions, both personal and those likely to be held by children, form an integral part of the audit process as students contemplate the cognitive demand posed by the subject matter. The SMLA forces the teacher to consider issues such as the use and interpretation of language and the nature of learning in science as they begin to translate the science curriculum for delivery in the classroom. As a consequence, the SMLA approach involves much more than simply identifying potentially difficult aspects of the science being taught, it encompasses both the nature of science and the nature of learning in science.

7.2.2 *The SMLA Process*

The SMLA approach has evolved from our research on how pre-service teachers learn in the physical sciences as illustrated in the preceding chapters of this book, and stems directly from student perceptions of their own learning. Figure 7.1 outlines the sequence of stages of the process.

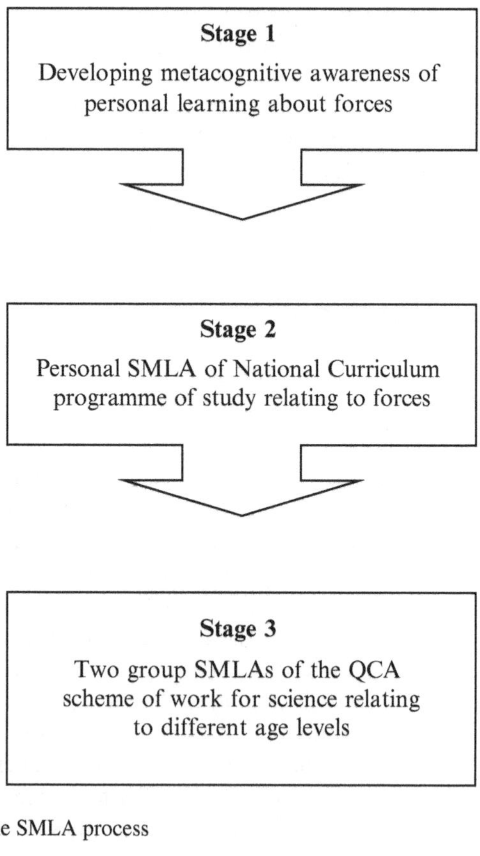

Fig. 7.1 Stages of the SMLA process

In Stage 1 metacognitive awareness of learning is developed during a university-based taught session on the particular subject. This provides the conceptual base from which curriculum requirements are considered through completion of an individual SMLA during Stage 2, and two subsequent group SMLAs during Stage 3. The Stage 3 SMLA engages the students in considering planning for children's learning within units of work drawn from national schemes of work for science. The three-stage process allows individuals to move from a position of learning at the personal level, where they develop insight into the nature of subject matter and the key ideas involved in developing qualitative explanation of it, to consideration of the curriculum and planning for children's learning at two different levels. The curriculum is examined firstly from an individual perspective in a general sense that is not context-bound (Stage 2) such that a sense of the progression of ideas at different levels is established prior to considering planning for teaching within a group context where ideas are shared, debated and negotiated. Having oriented thinking about the curriculum, the pre-service teachers are presented with a range of primary children's ideas prior to considering specific learning objectives drawn from a national scheme of work (Stage 3).

Whilst the conceptual domain of forces and motion present a particularly challenging area of learning for teachers (see Chapter 2), nevertheless the subject forms part of national curriculum requirements in several countries worldwide including New Zealand, Northern Ireland and China (CCCEA 2007; MoE, NZ 2007; MoE, PRC 2001). In countries without a specific national science curriculum, the topic is frequently attempted at the primary stage. In the USA, for instance, the National Science Education Standards for elementary children relate specifically to forces and motion (CSMEE 1996).

The Qualifications and Curriculum Authority's (QCA) scheme of work for primary science is adopted widely in English primary schools (DfEE/QCA 1998). The SMLA draws on objectives from two age-related units of work in order to provide a contrast in terms of subject knowledge cognitive demand. While learning objectives relevant to year 6 pupils (10–11 years olds) resonate strongly with several aspects of students teachers' own learning, and automatically precipitate in-depth discussion of pedagogical issues, the learning objectives for younger children (6–7 year olds) are not immediately obvious in a pedagogical sense and are likely to constitute a more challenging pedagogical task. This enables us to see whether emergent pedagogical insight is restricted to a narrow area of subject or whether general principles might be applied more widely in translating and interpreting the curriculum. In constructing the SMLAs, students not only attempt to identify explicitly the key conceptual framework, but additionally the related pedagogy of what is involved in the learning of the concepts. Integral to the process is the interpretation of the prescribed curriculum as a key analytical concept that requires intellectual engagement with science subject matter. This is important because there is evidence to suggest that inexperienced teachers who lack confidence in their subject matter knowledge are more likely to be limited in their interpretation of the prescribed curriculum (Harlen and Holroyd 1997).

Although the SMLA questions that provide the framework for analysing and translating the curriculum differ slightly, there is a good deal of commonality in that both require identification of:

- the key conceptual ideas to be developed by children
- particular challenges within the subject
- abstract or counterintuitive ideas
- language issues
- misconceptions
- factors influencing the learning process

The questions have been developed as a direct consequence of analysing student reflections of their own metacognitive processes and are derived from analysis of learner perspectives in several areas of study in the physical sciences as demonstrated in preceding chapters of this book. They are designed to facilitate learners' natural propensity to inform planning as a consequence of developing metacognitive awareness of their own learning.

7.3 A SMLA Case Study (Stage 1): Learning About Forces

7.3.1 The Participants

This case study presents empirical data documenting the learning of a group of three postgraduate students within the subject domain of forces as they undergo the SMLA process described. It involves three female students: Alex, Nasreen and Sarah, none of whom were science specialists, it having been some time since their formal science education. The group expressed no particular interest in science prior to embarking on the course and were anxious about the prospect of revisiting science and having to teach the subject in school. They were in their eighth week of a 1-year postgraduate course in primary education and, to date, had received only three university-based, 3 hour science sessions. They had no experience of planning science or teaching the subject in the primary setting at this point, although they had all spent time in primary schools in a voluntary capacity prior to the course. Their institution-based science education had so far involved them in identifying science in the primary education curriculum with a focus on scientific enquiry skills through practical exploration of vertical motion (using paper helicopters) and sound (using musical instruments and string telephones). Latterly they had explored forces within the contexts of floating and sinking, a simple arched bridge and a skydiver's descent as detailed in Chapter 2. In each context learners considered:

- The forces acting (including their direction and size)
- Balanced and unbalanced forces and motion

Teaching employed the metacognitive approaches described in detail in Chapter 6 and placed emphasis on the articulation of the learning process through critical reflection on experience and discussion of implications arising. In each context the intention was to develop qualitative explanation for the phenomenon in question. Critical review of personal learning constitutes the lens through which the student will examine the science curriculum in Stage 2.

7.3.2 Analysis of Prior Learning

The learning scenarios employed in developing knowledge and understanding of forces (floating and sinking, the simple arched bridge and the skydiver's descent) are described fully in Chapter 2.

7.3.2.1 Floating and Sinking

The group's learning was typical of that described previously for other learners in Chapter 2. A common initial response is to activate everyday knowledge of the phenomenon in making predictions about the behaviour of objects:

> I used my prior knowledge to determine if floating would occur but not my scientific understanding. (Alex)

Alex has a general sense that the material an object is made of will have influence and suspects that the size of the object is also significant. However, she expresses dissatisfaction with the incompleteness and coherence of her reasoning and admits to having difficulty in forming a 'satisfactory explanation'. Later, on reflection, she recognises that she has not considered the forces involved. She is able to pinpoint her misconceptions as 'the material the object was made of as the sole determining factor', and her failure 'to consider the upthrust produced by the water on the object'. She considers the forces view of the relative sizes of weight and upthrust, introduced by the tutor as being both counterintuitive and abstract, commenting that 'the ideas are alien and difficult to grasp'. Alex's response is typical of many learners in that forces are not incorporated into intuitive reasoning about floating and sinking and, although weight is often used in such explanations, it is usually conceptualised as a property of the object rather than a force.

Nasreen also thinks that the material an object is made of is an important determining factor in floating and sinking. She also expresses knowledge of density, defining this as 'how tightly packed the material is'. In reflecting on her predictions, Nasreen comes to realise that she held the misconception that 'weight alone is the determining factor' and that 'light objects float and heavy objects sink'. She had thought that all woods would float and all metals sink. Although she had associated knowledge of density, this was not being activated in reasoning about the phenomenon. Nasreen had clearly incorporated density into her schema relating to floating and sinking, but it remained a peripheral concept that was not located in the core knowledge used in explaining events. Both the weight of the object and its density appear to be strongly associated with the compositional material of the object and, although Nasreen clearly has knowledge of these concepts, she is unable to form the coherent view of their relationship needed to generate an explanation that she finds satisfying and useful. At one level Nasreen appears comfortable with a forces explanation based on the relative sizes of weight and upthrust and the notion of balanced and unbalanced forces, but on the other hand she experiences difficulty in developing a forces rationale in terms of action and reaction and struggles to rationalise that the water exerts an equal and opposite force on the object:

> I couldn't grasp the idea that the water 'knew' the size of the object (but obviously water doesn't consciously think!).

Sarah also reasons about floating and sinking initially using the premise that 'heavy things sink, light things float' and she combines this with knowledge that the material an object is made of is significant (in terms of 'heaviness'). Later she identifies this as a misconception and, although she finds the linking of weight with size (introduced by the tutor to support learning) a difficult bridging concept to access initially, once secure in this idea she finds it useful in reasoning about a number of contexts.

Although Sarah, Alex and Nasreen possess a considerable body of existing knowledge about floating and sinking, this knowledge is often incomplete or exists as isolated facets. The linking of weight with size provides a framework that helps learners rationalise this knowledge within a coherent explanation (see Chapter 2).

7.3.2.2 The Arched Bridge

Initially Alex thinks that a load placed on a simple arched bridge will push down on the bridge and that the bridge will push against its supports that, in turn, provide the 'push back to counteract the weight'. She argues that as the bridge fails to collapse when loaded, the supports must push back 'equally' and the forces must be balanced. On reflection Alex claims to have learnt that 'the weight of the load is transmitted through the structure' and is counterbalanced by the opposing force 'supplied by the supports'. While she seems confident in thinking of the bridge in terms of balanced forces, this is insufficient as a meaningful explanation:

> I saw it happening and therefore know it does, but because the force is invisible, it's difficult to understand the why of it all.

Although at one level Alex is satisfied with notions of the transmission and balancing of forces, she is still trying to rationalise this and has yet to arrive at what she sees as an appropriate causal explanation. This conceptual dilemma is exacerbated with the observation that the load can be increased further without the bridge collapsing:

> It's like the bridge knows how much to push back.

Nasreen also reasons that, as the bridge fails to collapse, the forces acting must be balanced. She states that she now understands that forces are being 'transferred' and 'shared' within structures, and describes the bridge as being 'only as strong as its supports'. The visual, tactile dimension had been critical in Nasreen's learning and she had begun to raise questions about other contexts such as the forces needed to open a door near to and further away from its hinges.

Sarah also claims to have developed a forces explanation for the 'transfer of force' through the structure and the 'balancing of forces by the bridge supports'. Like Nasreen the physical experience of forces had been a critical feature of her learning experience. She identifies a significant difficulty in developing understanding as:

> I couldn't 'see' the force – just its effect – knowing this helps my understanding a great deal.

The latter represents important pedagogical knowledge. Not only does Sarah recognise that the inability to 'see' force acting and being transmitted as a key difficulty in generating explanation, she also acknowledges that the *realisation* of this is an important element of her own pedagogical understanding.

7.3.2.3 The Skydiver's Descent

Alex's initial conception is that a parachute is needed to alter the balance of the forces acting on the skydiver in order to slow the descent. She has become comfortable in employing a forces perspective and identifies the relevant forces as 'air resistance' and 'gravity'. However, on reflection, she realises that she has not considered that forces would become balanced in the final descent when the parachute is open and terminal velocity reached. In terms of learning difficulty, Alex finds the lack of physical experience of the phenomenon limiting, making it difficult to recognise changes in motion such as acceleration, deceleration and constant velocity. However, at this stage she seems confident in employing abstract reasoning about the relative sizes of forces and regards this scenario 'the easiest to deal with because of its logical nature'.

The role of the parachute in slowing the skydiver's descent is also the initial starting point for Nasreen's reasoning, however, she makes no reference to the forces involved or their relative sizes. Nasreen recognises that she has not considered the issue of balanced forces acting at constant velocity and struggles to rationalise that this could occur both before and after the parachute opens during the descent. Her intuitive view was that the skydiver would continue to accelerate indefinitely with the parachute open and she became aware of the difficulty in accepting the theoretical notion of balanced forces and terminal velocity as it militated against her common-sense reasoning and everyday observation of falling objects. Nasreen points to the potential confusion in differentiating force and motion in teaching and learning and begins to raise interesting conceptual questions such as whether terminal velocity is always the same irrespective of the weight of a person.

Sarah has a similar starting point to Nasreen viewing parachutes as a means of 'slowing the rate of fall' and identifying her 'misconception' as 'a lack of consideration of forces acting and their relative sizes'. She reports difficulties encountered in learning as recognising forces acting together with their attendant effects on motion as they come in and out of balance as the descent progresses and the parachute is opened. For Sarah, the idea that forces can become balanced and velocity constant both before and after the parachute opens is problematic.

7.4 A SMLA Case Study (Stage 2): The Individual National Curriculum SMLA

Having developed awareness of the challenges of personal learning, the next step in the SMLA process was to use this knowledge as a metacognitive lens in considering the subject matter to be taught to primary children. Stage 2 of the SMLA process involves critical scrutiny of the national curriculum programmes of study relating to forces (Fig. 7.2) covering Key Stage 1 (5–7year olds), Key Stage 2 (7–11 year olds) and Key Stage 3 (11–14 year olds).

In meeting the needs of the more able primary child, it is useful to have knowledge of the Key Stage 3 requirements. The SMLA used to guide and focus student thinking encompassed the following questions:

1. What key ideas are being developed within the programme of study?
2. What ideas do you think might be particularly challenging for learners?

Key stage 1 (5-7 years)	**Key stage 2** (7-11years)	**Key Stage 3** (11-14years)
Children should be taught about:		
The movement of familiar things such as cars going faster, slowing down, changing direction.	The direction in which forces act.	How to determine the speed of a moving object and to use the quantitative relationship between speed, distance and time.
Both pushes and pulls are examples of forces.	That friction, including air resistance, is a force that slows moving objects and may prevent objects from moving.	Ways in which frictional forces, including air resistance, affect motion.
When things speed up, slow down or change direction, there is a cause such as a push or a pull.	That objects are pulled downwards because of the gravitational attraction between them and the Earth.	The weight of an object on Earth is the result of gravitational attraction between its mass and that of the Earth.
	That when objects such as a spring or table are pushed or pulled, an opposing push or pull can be felt.	That unbalanced forces change the speed, or direction of movement of objects but balanced forces do not.

Fig. 7.2 Some key ideas about forces developed within the English National Curriculum

3. Are any of these ideas counterintuitive or abstract?
4. Did you have any personal misconceptions in relation to this subject matter?
5. Are any language issues likely to arise in the learning of this subject matter?
6. What factors helped and/or hindered your own learning in this area?

7.4.1 Key Ideas Within the Programmes of Study

The key ideas identified by the group are presented in Fig. 7.3a. Broadly speaking, these ideas address: what forces *are* (for example, pushes and pulls, weight, gravity, friction), how forces *act* (in pairs, on all objects, transmitted within structures, in balance or unbalanced), what *effect* forces have (in terms of motion, direction) and how forces act in *particular contexts* (for example, floating and sinking).

The main focus of individual reflection concerned the concept of balanced and unbalanced forces and their consequent effects on motion. This is unsurprising in that it constituted a key element of students' own learning in forming a central idea that had enabled them to rationalise observations in a range of contexts. In scrutinising the curriculum, most attention was paid to the culmination of learning relating to the issue of balanced forces in the Key Stage 3 programme of study, and less attention was paid to the formation of key concepts in the earlier stages. Sarah is the only person to identify pushes, pulls and frictional forces as early key ideas.

Students' learning experience of subject matter is, clearly, a major influence in their interpretation of the curriculum. Critical elements of Alex's learning about forces had been the existence of an upthrust exerted by the water and the development of a forces view of floating and sinking. In considering key ideas developed within the curriculum she identifies 'weight for size' and the 'upward thrust of water' even though these concepts are not mentioned explicitly within the documentation. Similarly there is no reference to the 'transmission' of forces; a key idea emerging for Alex from the arched bridge context. Nasreen's struggle with the concept that weight per se is not the determining factor in floating and sinking (a key feature of her learning) is reflected in her inclusion of the notion of weight for size as a key curriculum idea. Also testament to this is her point pertaining to 'the more something weighs the more gravity is acting on it'; again an idea stemming directly from her own learning experience. Less obvious is how Sarah's view of the curriculum resonates with her own learning and she demonstrates a more detached and balanced view. This is, perhaps, related to the key recognition in her own learning of the need to differentiate forces and their attendant effects on motion as they come in and out of balance in explaining movements in different contexts. Sarah was much more confident in both starting points and subsequent learning and was able to look more critically at the curriculum, being not so contextually bound by her own conceptual difficulties. It seems that personal learning, especially that entailing conceptual conflict and struggle to develop causal explanation, has impact upon the way in which curriculum requirements are interpreted.

	Alex	Nasreen	Sarah
a. key ideas	• forces come in pairs • forces act on things • weight is a force • forces can be transmitted • when forces are balanced things stay the same • when forces are unbalanced, objects slow down, speed up, change direction • weight for size • upward thrust of water as a force	• forces acting on everything • balanced forces act on stationary objects • unbalanced forces cause changes in movement • how the forces are balanced affects the speed • the more something weighs, the more gravity is acting on it • weight for size	• pushes and pulls are both forces • gravity as a force pulls objects downwards • frictional forces (air resistance) influence motion • the movement of an object does not change when forces are balanced • unbalanced forces change the speed or direction of movement of an object
b. challenging ideas	• weight for size means an enormous object can float • the notion of the transmission of invisible force • a stationary object has forces acting on it • forces are balanced even when an object is travelling at a fast but constant speed (terminal velocity)	• large, heavy objects can float and small, light objects can sink • a stationary object still has forces acting on it	• more than one force can act on an object at one time • understanding balanced/unbalanced forces and their effects on an object • knowing a little about forces but really understanding it
c. counter-intuitive ideas	• weight for size • forces being balanced even when an object is travelling at very fast constant speed	• more gravity works on objects with a larger mass • weight as a force, not heaviness	• forces being balanced/unbalanced

Fig. 7.3 Individual National Curriculum SMLA

7.4.2 *Challenging Ideas*

Individuals identified a range of challenging ideas within the programmes of study (Fig. 7.3b) and these were essentially similar in nature to those encountered in their personal learning. The universality of forces acting, particularly where this is imperceptible in stationary objects, formed a central concern. Some had found balanced and unbalanced forces and their effects problematic and the notion of weight as a force was a difficult conception for all. Although the curriculum relating to forces does not specify contexts, in articulating challenging ideas, both Alex and Nasreen employed contextually bound ideas and only Sarah interpreted the curriculum in a context-independent way. Nasreen and Alex considered the visual imperceptibility of forces a problem in explaining the transmission of force within static structures. There is a tension between what students view as significant in personal learning, and the context-independent concepts developed within the curriculum. Nasreen and Alex find it difficult to extract such generalisations from contexts.

7.4.3 *Abstract or Counterintuitive Ideas*

Having recognised aspects of subject that are likely to present particular difficulty in learning, the next stage of the SMLA is to focus student thinking on why this might be so. Our research has highlighted that pre-service teachers consistently identify the abstract or counterintuitive nature of physics concepts as a conceptual hurdle in learning and consequently, the SMLA asks them to anticipate and consider any such aspects within the prescribed curriculum (Fig. 7.3c).

Nasreen pointed to 'weight as a force' as a likely central concern in all the contexts and cited it as a particularly counterintuitive and abstract notion:

> The fact that things can weigh different amounts on the moon, in water etc. is very strange because what's making them up hasn't changed.

Nasreen's struggle to rationalise weight as a consequence of gravitational attraction derives from her conception of gravity. She sees gravity as an independent entity emanating from the Earth and acting on all objects equally:

> The idea that more gravity works on objects with a larger mass – so would something weigh nothing if there was no gravity there, but still the same mass of the object? But surely it just comes from the Earth and pulls everything towards it the same?

Alex is also still in the process of assimilating the notion of 'weight for size' in explaining floating and sinking. The problem lies in the fact that intuitively weight appears to be the sole determining factor and that although on the one hand students are often able to offer explanation based on density or an association of factors (see Chapter 2), when it comes to making predictions they are likely to revert to weight as the sole influence. The conceptual challenge of linking two concepts (weight and size) in formulating an explanation is exemplified in Alex identifying this as counterintuitive (Fig. 7.3c).

All three learners reported the notion of balanced/unbalanced forces to be contrary to intuitive thinking. In intuitive thinking balance implies that an object is stationary and not moving.

> That forces are balanced when an object is travelling at very fast constant speed just doesn't seem right. When something is balanced you tend to think it's still not moving and the skydiver it can be balanced before and after the parachute is open – at two different speeds – that's odd.

7.4.4 *Personal Misconceptions*

Students' views of what constitutes 'misconception' may differ from the way in which educational psychology and science education employ the term to talk about learners' misunderstandings of scientific concepts. Alex, Nasreen and Sarah applied a broader, less-defined view. Sometimes this encompassed sources of uncertainty, such as where Sarah pinpoints as misconception her 'confusion' in regard to mass and weight and 'lack of clarity' in regard to gravity. She is not explicit about her conceptions of these notions; it is possible that she is unable at this point to articulate them further. Learners also use 'misconception' to denote aspects of subject they find difficult to understand such as when Nasreen points to her cognitive struggle to rationalise how the water 'knew' the size of the object in exerting an opposing force. Alex raises the issue of 'accepting' explanation and 'understanding' it in regard to the transmission of forces within bridges:

> I knew bridges had supports but never understood the notion of the transmission of weight – just accepted that it must happen.

They also cite as 'misconception' knowledge or ideas that they had not considered prior to a learning activity:

> Floating and sinking – I never thought material and mass were determining factors. (Alex)

At other times, students are much clearer about the nature of their misconceptions:

> I thought all wooden objects would float and all metal objects would sink. (Nasreen)

Exploring personal misconceptions in relation to curriculum subject matter serves the purpose of alerting learners to the power of personal constructs in learning as well as highlighting the differences between personal misconception and scientific explanation. They become aware of how robust intuitive ideas can be and how difficult they can be to influence in teaching.

> I know it's not the full story, but something inside me tells me it's the weight that makes it sink. (Alex)

7.4.5 Language Issues

All three students raised the issue of language and, in particular, the understanding of scientific terms such as weight, mass, gravity and density as being very important in developing their own understanding. They became aware of the potential for confusion where use was imprecise and definition vague:

> I became aware and a little concerned about the actual meaning of some words and the difference between them (such as weight and mass). (Sarah)

> Weight is used in everyday language to mean mass, not force. (Alex)

The conflict between the everyday uses of language, different personal interpretations of words and expressions, and scientific definition is acutely apparent in the context of forces associated language. A classic example of this is provided by the term 'weight'. In everyday usage this is usually interpreted as the 'heaviness' of an object, it frequently pertains to particularly heavy objects and implies gravitas, burden and mass. In science, however, weight is a force defined in terms of gravitational attraction and this is directly at odds with the intuitive view of weight as a property of an object as we have seen in Chapter 2 of this book. The recognition of this constitutes important pedagogical knowledge:

> I think where the difficulty lies [*with 'weight'*] is that it just goes against everything they know – it's the gravitational pull, the gravitational force – so there's a language issue and a counter-intuitive idea – for everyone really. Why don't they just call it mass –that's what we read on a scale – I think that's why we have trouble with it – why it's hard to understand. (Nasreen)

7.4.6 Other Factors Influencing Learning

In developing pedagogic knowledge it is important to consider not only the subject matter per se, but also other factors that influence the learning process. In identifying positive influences on individual learning, all three teachers readily acknowledged the important role of practical demonstration and investigation:

> The practical demonstration of the bridges bearing loads and witnessing the supports taking the weight and feeling the strain on holding the rod [*a cantilever bridge model*] was very important for me. (Alex)

> In the cantilever structure I could actually feel the weight being shared. (Nasreen)

An overwhelming influence is that of having tactile, visual learning experience and the opportunity to engage in the discussion and testing of personal ideas. The group made less comment in regard to factors hindering the process. However, Alex provides insightful observation:

> I am still to some extent hindered by the complexity and enormity of the concepts involved.

Although elements facilitating learning, such as the value of experimental work, and the role of discussion in learning science, are well-documented in science education literature, for teachers in the initial stages of professional development they constitute significant personal insight. In addition, other insights were subtle and discerning:

> I have learnt that in order to understand you have to question your knowledge. Knowledge and understanding are put together but they certainly don't come together or mean the same thing. (Sarah)

7.5 A SMLA Case Study (Stage 3): Scheme of Work SMLA

Having oriented thinking about the curriculum through the previous SMLA process, Stage 3 sought to explore how this developing metacognitive awareness might impact upon pedagogical awareness as students contemplate the implications for teaching about forces. Prior to this they were introduced to some primary children's ideas regarding forces that have been documents by the SPACE research project (Russell et al. 1998). These included common misconceptions such as:

- objects move naturally without cause
- objects move because of their features such as possessing wheels or some means of propulsion such as oars or wings
- forces are not involved when objects are moving (for example falling) or when objects are stationary
- objects fall because they are heavy
- forces belong to objects and reside within them
- forces are used up in motion
- gravity emanates from the Earth and attracts objects like a magnet

The intention here was not to provide an exhaustive list of children's likely misconceptions in the field, but rather to alert the teachers to some common constructs and to encourage them to think further in the light of their own learning.

7.5.1 Group SMLA of QCA Unit 6E

As part of the QCA Scheme of Work for Science, Unit 6E (Forces in action), children aged 10–11years are taught that:

- the Earth and objects are pulled towards each other
- this gravitational attraction causes objects to have weight
- weight is a force

Chapter 2 demonstrated the inherent difficulty presented by these ideas and Nasreen, Alex and Sarah had identified them as challenging notions in their own learning. Consequently, in planning for children's learning the group are faced with structuring learning that they have personally found problematic. In order to support them in considering pedagogical implications, the following SMLA questions were used to stimulate discussion:

- what do the children need to understand in order to make sense of the learning objectives?
- are there any potentially challenging areas in this content?
- are any counterintuitive or abstract ideas involved?
- are there any language issues that might arise?
- what ideas are the children likely to have in this area?
- are there any other important points that need to be considered in the teaching and learning of this subject matter?

The Stage 3 SMLA employs the knowledge of the whole group as a stimulus to discussion, creating a forum in which ideas are shared, considered, refined and debated.

7.5.1.1 What Will Children Need to Understand in Order to Make Sense of the Learning Objectives?

This question requires students to determine key ideas children will have to possess or develop in building meaningful understanding. The group's response is summarised in Fig. 7.4a.

Although one might argue that the points generated are ill-defined and pedagogically naïve, it must be remembered that they constitute *embryonic* steps in developing

---

Group A Unit 6E SMLA (Forces in action)

---

a. What do the children need to understand?

- the difference between weight and mass
- gravity is a downward pull to the centre of the earth and that objects are therefore pulled towards the earth
- heavier objects have more weight therefore more gravitational pull
- everything has force acting on it

---

b. Are there any possible challenging areas in this content?

- the confusion between weight and mass because weight is being used incorrectly in everyday language
- forces are acting on objects even when they are still - how do you know ?

---

c. Are there any counterintuitive ideas involved?

- you can't actually see forces – they are an abstract idea
- weight and mass

Fig. 7.4 Group responses to QCA Unit 6E SMLA

professional insight. Furthermore, they are generated in the early stages of professional education by individuals who have not yet had the experience of planning and teaching science lessons in school.

As a starting point the group decided it would be important to reinforce the concept of forces as pushes and pulls and to identify forces acting, including their direction. It was important to identify weight as a downward force in a range of contexts. An integral part of developing the concept of weight as a force was the need to understand gravitational attraction. Although the latter sounds vague and to a degree obvious, transcripts of group discussion reveal that these conclusions were arrived at as a result of in-depth thinking that involved examination of their own conceptual struggles and careful consideration of pedagogical implications for children's learning. Alex, Sarah and Nasreen became concerned about needing to help children differentiate weight and mass:

> Nasreen: 'Cos when you stand on the bathroom scales that's not your weight, it measures your weight and converts it to kilos, so you get a reading of mass.
> Alex: Oh that's really complicated, but they would need to understand that what they call weight isn't wouldn't they?

They are forced to contemplate what constitutes appropriate knowledge for children of this age and, in doing so they critically examine curriculum intentions:

> Nasreen: But I think that's too hard for Year 6, is that what the NC really asks for? Even when we looked at that, the concepts of weight and mass, it was really difficult.
> Sarah: I think we need to look at what they know about, like weightlessness in space and moving about on the Moon. We could show them videos and use the internet to get a satellite view and the film of men landing on the Moon.

Sarah's comment about needing to know about 'what the children know about' indicates her growing awareness of the need for teaching based on the learner's needs and she is already contemplating resources appropriate for her purpose to create meaningful learning contexts of relevance to children.

The group decide that important knowledge is that objects 'have weight' because of gravitational attraction between them and the Earth. In writing and conversation, they use the terms 'gravity' and 'gravitational attraction' interchangeably and make no attempt to differentiate between the two. They discuss the need for children to understand that 'the heavier the object, the more gravitational force is acting' and that they would need to 'make sense' of the concept of weight as a force. They recognised that the latter was not without difficulty and it was anticipated that weight and mass might need to be differentiated, although this is not stated explicitly in the unit learning objectives:

> Alex: The concepts of weight and mass make this all so hard.
> Nasreen: Yes, I find it difficult that more gravity acts on bigger objects – which is right –isn't it?
> Sarah: If you think about it they have to use more force to pick up a bigger object – so it's kind of the same and I don't think they'd find it difficult that more gravity acts on bigger objects.

Alex: Yes that's a really good way of looking at it – it pulls everything downward and we have to push against it to pick something up.
Sarah: In a way looking at space is helpful by way of providing a contrast – 'cos you've got spacemen floating round in the air – the Moon is always difficult – that's no gravity.
Nasreen: So gravity is pulling you down to the floor and on the Moon it's not?
Sarah: But you don't weigh nothing on the Moon – you weigh less on smaller planets – isn't it cos the Moon's 1/6th the size of the Earth and so the gravitational pull is 1/6th as well –so you weigh less?
Alex: If that's what you focus on for a whole lesson I think that would work OK.

This discussion represents a process of negotiation of curriculum meaning in which complex concepts are broken down and appropriate learning and learning contexts identified. This approach is radically different in nature from one predicated on simply deriving learning objectives from schemes of work or the curriculum uncritically.

7.5.1.2 Are There Any Potentially Challenging Areas?

Weight as a force, with the attendant potential for confusion in discerning weight and mass, was quickly identified as a very challenging concept and its appropriateness for children of this age was questioned by the group (Fig. 7.4b). It presented a cognitive and pedagogical dilemma in that it was only through the process of differentiating weight and mass that they had arrived at a personal understanding of weight as a force, yet the learning objectives make no reference to mass. The difficulty involved in this should not be underestimated; the translation of the concept 'weight is a force' into meaningful teaching and learning is a core difficulty for learners of all ages in understanding basic Newtonian mechanics (see Chapter 2 for a fuller discussion). Although this is a curriculum concept appropriate for the 11–14 year old (Fig. 7.2), it is often addressed in the later primary years and needs to be considered by primary teachers. The translation of this into effective teaching is not without difficulty and this formed a focus of debate:

Nasreen: I can't really understand how I would weigh less on the Moon 'cos I'm still the same shape and size – so in my head I should still weigh the same – so what am I going to say when they [*the children*] say that to me?
Alex: We have to make it clear that it's a force – not heaviness – they need some kind of answer as to what weight is – I think that can only come through feeling the force.
Sarah: For children we'd need simpler ideas and experiences like feeling different weights and dropping different masses, like feeling the difference between suspending a brick in the air and then in water – things like that would help.
Nasreen: Yes they might not all get it but they would have the experience and that would give a foundation for later.
Alex: I think that introducing floating and sinking would confuse them all the more – just focusing on lifting things and dropping things would do – the more you have to lift, the more gravitational pull – so you just need to feel it.
Sarah: We could think about what it would be like if there was no gravitational pull – everything would just float off and we could use the Moon for comparison.

Another challenging area identified by the group was that forces 'act on' objects as opposed to 'belong to' them or 'reside within' them. They thought this might be addressed through having a focus on physical experience of pushes and pulls in a range of different contexts followed by looking more closely at forces at a distance such as magnetic attraction and repulsion and gravitational pull. A particularly important part of this would be to provide learners with the experience of physical forces acting in stationary structures and contexts. The group is becoming aware of the need to structure learning experience and not to be ambitious in covering too may learning objectives at once:

Alex: We could use diagrams and arrows to show that gravity comes from the centre of the Earth. The direction's really important.
Nasreen: But you don't really think like that – it's too abstract – you just feel it.
Sarah: Yes you have to visualise it.
Alex: The Earth attracts/pulls objects towards it.
Sarah: The Earth isn't magnetic, it's not magnetic attraction. Gravitational attraction causes the objects to have weight, the pull depends on the proximity of the objects – should we go into that?
Nasreen: I don't know I think that would be far too much – we should just stick with the idea that heavier things have more weight.
Alex: We mustn't have too may concepts at once – if I cover too much, I just get confused – we need logical stages.
Sarah: Yes like in floating and sinking how we put weight and size together to explain things.

The discussion represents a cycle of refinement in which discourse enables students to negotiate an appropriate curriculum through critical scrutiny of subject matter from the learning perspective. It places the focus on learning and learners as constructors of meaning, as opposed to teaching as a process of transmission, a view commonly subscribed to by the novice practitioner.

7.5.1.3 Are there any abstract or counterintuitive ideas involved?

In order to place focus on the reasons why subject matter might be difficult for learners, the group considered potentially abstract or counterintuitive ideas inherent in the content. The fact that forces cannot be seen directly was quickly recognised as an abstract and potentially counterintuitive notion (Fig. 7.4c).

Nasreen: They can't see forces, so they wouldn't think of forces acting on stationary objects would they?
Sarah: They might think that if something was still on the ground, like a house, there are no forces acting on it.
Nasreen: They'll think it's the foundations making it attached as you can't lift it up and move it.

This discussion evolved into a focus on two basic ideas: first, the notion that forces act on objects and do not reside in them and second, that forces cannot be seen, only their effect is manifest. In essence this is the same problem as experienced in learning about light and electricity, where the only the manifestation of the phenomenon is accessible to the learner and therefore teaching needs to find a way to make the imperceptible perceptible in order to address what might be termed a 'black box' element that can act to hinder learning. There is no easy solution to this conceptual dilemma in teaching forces to primary children and the students concluded that the best way to address it would be to provide physical experience of applying forces to change the motion of objects so as to focus clearly on action and effect:

Alex: Maybe we start with the idea that everything has forces acting on it.
Nasreen: But if the object's still and gravity's acting downwards, what's pushing it up to keep it still?
Alex: It's the table's resistance or something.
Nasreen: But it's just giving it something to rest on – not pushing it back.
Alex: I don't think you'd introduce that to a year 6 class – it's too abstract.
Nasreen: Perhaps we could just get them to feel the forces – like when we used a ruler and string to make the cantilever bridge and could feel the strain- perhaps they could measure the pull?
Alex: Or we could use big springs to get the same effect.

7.5.1.4 Are There Any Language Issues Involved?

As differentiating the terms *gravity*, *weight* and *mass* had been a major part of their own learning, and it naturally formed a focus of the discussion of children's learning. The group thought that children would probably resort to everyday language based on 'push' and 'pull' in discussing movement and may not readily engage in applying specific 'forces' terminology. Consequently, developing scientific language would have to be a key feature of teaching:

Sarah: I think pushes and pulls are hard because when you're picking something up or doing something, you don't think of pushes and pulls you think of strength.
Alex: The language's a nightmare if you think about it.
Sarah: Yes we'd really have to know about how the children were describing things.
Alex: Yes and we'd have to think about how we're describing things ourselves.

7.5.1.5 What Are Children's Likely Ideas Likely to Be?

In considering examples of children's ideas (Russell et al. 1998) and combining these with their own experiences of children and personal learning about forces, a variety of possible conceptions were anticipated. Children's possible conceptions of gravity were a key issue in discussion and students began to recognise their potential significance in learning. The anticipated misconceptions ranged from children having 'no real' understanding of the concept of gravity to thinking that gravity will operate equally on all objects irrespective of mass. This acted to reinforce the issue of understanding the learner's perspective and, in particular, what learners understand by the language they employ and the language used by teachers in teaching.

7.5.1.6 Are There Any Other Important Points That Need to Be Considered in the Teaching and Learning of This Subject Matter?

Conclusions regarding effective practice concerned pedagogical issues ranging from the use of appropriate resources in teaching such as film and photographs taken from space, to the need to provide opportunity for learners to express and discuss their ideas within lessons and to participate in investigations in order to gain concrete, physical experience of the abstract concept of force. There were clear implications for their own subject knowledge and they became critically aware of the need for focused 'teaching points' and the need for logical stages to be discerned in order to scaffold thinking in helping to navigate learners through the subject matter effectively. They recognised that language, communication and confusion over scientific terminology may form barriers to learning about forces.

7.5.2 Group SMLA of the QCA Unit 2E (Forces and Movement)

Unit 6E learning objectives had resonated strongly with personal learning experiences of forces and the SMLA demonstrates that they were able to utilise metacognitive awareness effectively in anticipating teaching and learning. The next stage involved scrutiny of the following learning objective drawn from unit 2E (Forces and movement) designed for 6–7 year olds:

7.5.2.1 That Pushes and Pulls Can Make Things Speed Up or Slow Down or Change Direction

In examining what seems like an obvious statement, our intention is to ascertain whether students are able to apply the same depth of thinking to pedagogical implications as they had previously. Would they be able to utilise their metacognitive awareness of learning in planning for teaching objectives that did not resonate so strongly with their own learning experience? The format of the SMLA was the same as that described for Unit 6E and group responses are summarised in Fig. 7.5.

The following discussion makes no attempt to analyse outcomes in relation to the specific SMLA questions posed for Unit 2E as the discussion that arose was more fluid and less structured than previously as students became more confident in professional discourse.

From the outset, the group were cognisant of the conceptual challenge posed by the learning objective and this is reflected in their speculation about what young children might think in this area:

> They're just going to think that things come naturally and they don't see any need for forces and anything like that so their initial response will be 'it just happens and that's it' and that it's meant to be like that – 'cos that's what it does and that's what it feels like. (Alex)

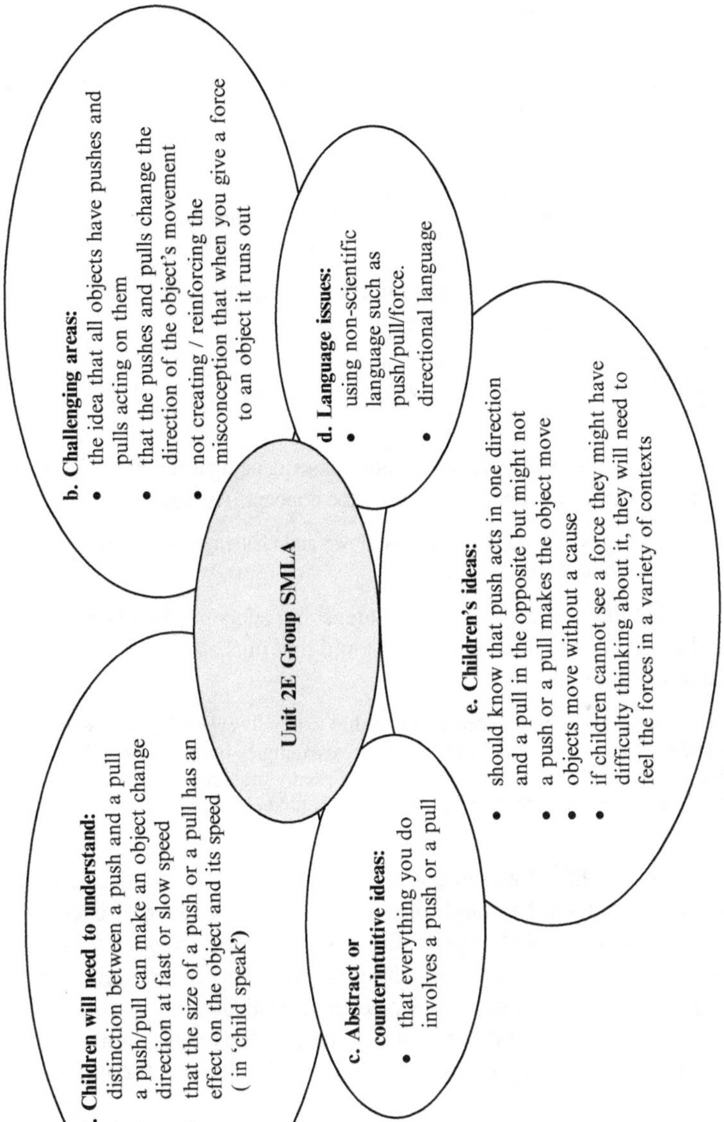

Fig. 7.5 Group response to QCA Unit 2E SMLA

Generally, they thought that children may regard movement as a 'natural' event without there being a need for a cause. In identifying what children will have to understand in order to meet the objective, the consensus was that a sensible starting point would be to develop meaning for the terms 'push' and 'pull' through concrete experience involving everyday contexts:

> Alex: On the face of it it's a very simple statement I think, the difficulty is in figuring out what children will have to understand in this area and seeing it from their point of view. What do you think they might not be recognising?
> Sarah: We quite often don't recognise pushes and pulls in everyday movements and I'm thinking that what we'll have to do when teaching is ensure that they do recognise and explore pushes and pulls and detect them in everyday experience. You can think about that in an abstract way immediately but young children can't.

They readily recognised that young children are unlikely to be unable to reason effectively without concrete experience:

> Nasreen: It needs to be visual and kinaesthetic.
> Sarah: Yes, practical – we need to get them all standing up and pushing and pulling – feeling the forces.

The group became aware of the need to consider sequencing teaching logically in order to help children to build their understanding of the concept. As Nasreen observed:

> The challenge for us is how do we break that down and challenge that in our teaching so as to make our teaching effective.

They decided that the next logical step in a teaching sequence would be the linking of cause and effect such that the children should link pushes and pulls with changes in movement and direction:

> Alex: The idea that everything is caused by pushes and pulls even when you write you're pushing and pulling – when you throw a ball – give examples from everyday things.
> Sarah: We need to let them see that these are connected – the force and movement.
> Nasreen: A push or pull can change direction – that could be easily shown couldn't it – it's not a challenging area.

Following on from this it was important to link the size of pushes and pulls with effect. This could be taught through physical, tactile experience of everyday contexts such as movements of the body, playground settings, toys and games. Also they would have to ensure that young children gain experience of contexts where pushes and pulls are not immediately obvious in activities such as writing or eating. While the learning objective did not explicitly require the recognition of forces acting on stationary objects, they thought that this would be too difficult and abstract a concept for young children:

> Sarah: They would have to understand about gravity and perhaps friction.
> Nasreen: Yes and transmission of forces in bridges etc. – that would be far too hard.
> Alex: But it doesn't say we have to teach that, does it – it just talks about speeding things up and slowing things down or changing direction.

A key element of discussion centred on the effect of friction in influencing movement. In thinking about what happens as a toy car is pushed and moves across a carpet, slows down and eventually stops as a result of frictional force, there ensued a pedagogical dilemma. It was anticipated that children would simply view such changes as occurring 'naturally' and see no need for an explanation based on forces (Fig. 7.5e). Also, the notion of pulls and pushes being consumed may possibly lead towards the development of a misconception:

Alex: It's difficult because if something's moving on a carpet like a toy car that's been pushed, it slows down and eventually stops, it's caused by friction but the children aren't going to realise that are they? It's too hard at this stage.
Sarah: They'll think that's not the force, they'll just think that's your' push' running out.
Nasreen: So they'll think that once you've pushed something you've then stopped acting on it – your force has stopped acting on that thing – so it'll work for a bit then it'll stop then it'll need more push.

It was generally felt that thinking about everyday movements in terms of pushes and pulls was not a natural way of thinking for young children. The issue of explaining the slowing down and stopping of a moving toy car without resorting to inaccessible terms and not seeking to embed the misconception of consumable forces became a central concern:

Alex: Yes that's hard for us, we've learnt an awful lot ourselves but were not actually communicating all that knowledge, it's a case of preparing the children for where they are.
Nasreen: But what would you say slowed it down, why? Cos you're saying that pushes and pulls can make things speed up and slow down.
Alex: So would you introduce friction?
Sarah: I think I wouldn't there'd be no point, it's too hard. I think I'd just notice it and tolerate any questions, cos they don't do friction until much later so what could you say?
Alex: Yes they might not even ask why but we need to think about what we'll do before so we don't get confused or confuse them in our teaching. We could just get them to feel how hard it is to push against different surfaces and not call it friction.

This is a mute point and one to which there is no convincing answer. The activities teachers employ in teaching young children about forces typically involve identifying pushes and pulls in a variety of contexts and the linking of pushes and pulls with subsequent changes in movement. In one sense this serves to embed the misconception of force as a consumable commodity and there is no easy solution to this pedagogical problem. Having considered the inherent difficulties, Sarah's conclusion seems the sensible, if not the only realistic approach that could be taken.

The group also thought that there was probably a great deal of work to be done in developing children's language skills at this stage:

Sarah: There's a big language issue in this one – we need to think about the language that they might need to develop in order to understand this.
Nasreen: The idea of pushes and pulls?
Sarah: Would you call pushes and pulls forces at this stage?
Alex: I think we'd stick with pushes and pulls, you might mention it – forces – but just carry on talking about pushes and pulls.
Nasreen: There's no point in telling them if we're not going to use it.

The group concluded that 'push' and 'pull' need to be defined through concrete learning experiences in real-life contexts in order to develop *real* meaning for the terms. While it was likely that children would understand the meaning of ‘push’, the group were less sure of their ability to distinguish ‘pull’ in complex movement and thought this would shape the attention-focusing questioning they would use in teaching. Furthermore, they questioned whether children would find it easy to make the connection between push/pull and movement, or change in direction, but generalising across different contexts was more problematic as they may not recognise pushes and pulls acting in many everyday contexts. A main point of contention lay in the use of the term ‘force’. It was envisaged that children might depend more on push and pull as ‘force’ was not used naturally in everyday language in this way. Thus, teaching would have to introduce this term carefully within a meaningful framework.

7.5.2.2 Other Important Aspects of Learning

Reflections concerned what the group perceived as contributing towards effective practice. They subscribed unequivocally to teaching based on visual, tactile, concrete experiences within which young children might embed the notion of pushes and pulls. Richness, variety and choice of experience were important and learning experiences must be related to real-life contexts and provide ample opportunity for learners to discuss their ideas and observations. Teachers would have to use attention-focusing questions to help children make appropriate observations to link cause and effect, without this learning would lack meaning. They thought it important to make explicit experience derived from investigations and there was much discussion of locating learning experiences within the context of play. Covering the subject matter in logical steps and focusing on key vocabulary were critically important elements of their approach.

7.6 Discussion and Implications for Teacher Education

Research has indicated that experienced teachers operate within a conceptual framework of pedagogy that integrates coherently their knowledge and beliefs about the nature of science and subject matter in relation to children’s learning. Pre-service teachers may, however, experience conflict between personal views of teaching and learning and classroom practice (Brickhouse 1990; Brickhouse and Bodner 1992; Lederman 1999) and their personal histories of learning science are known to have significant influence on how they teach science to children (van Zee and Roberts 2001).

It, therefore, appears that the instructional strategies of beginning teachers are derived from personal experience of learning and development is strongly influenced by contextual and individual factors including epistemological beliefs and personal goals (Adams and Krockover 1997). This often results in teachers adopting institutional conventions that emphasise procedures instead of developing strategies based on promoting understanding of science concepts. Calderhead (1996), for instance, argues that student teachers are likely to draw upon observed practices of supervising teachers rather than their own knowledge base, and found that they gave low priority to subject knowledge in planning. Halim and Meerah (2002), in a study of pre-service teachers' PCK relating to physics concepts, found that most teachers in their study were unaware of children's likely misconceptions, or were unable to anticipate them or offer detailed teaching strategies to address them in practice. Their main approach to teaching was one of restating the principles involved, that is they viewed teaching as 'telling' as opposed to 'representing' subject matter (Gunstone et al. 1993). Halim and Meerah concluded that a strong focus on developing PCK at pre-service level would be beneficial because, although teaching experience over time provides some opportunity to eradicate teacher misconceptions and develop PCK, the process is idiosyncratic and context-dependent.

7.6.1 What Can the SMLA Approach Contribute to Teacher Education?

The case study described above constitutes powerful evidence that given appropriate circumstances, pre-service teachers, even in the early stages of professional development, can interpret the curriculum critically such that important pedagogical implications arise for the planning of teaching. The intellectual conduit that channels their thinking is created within the act of developing metacognitive awareness of their own learning. When such awareness is applied to interpreting the school curriculum, the case study shows that the students control and regulate planning through the filter of their own learning experience. This process is not just confined to parts of the subject matter that resonate strongly with their own conceptual journey, as might be anticipated, but knowledge and skills are transferable to subject matter where the conceptual demand is different, in this case ostensibly much less demanding. This is evidenced in the group's discussion of the possibility of creating or reinforcing the misconception of forces residing within objects when a toy car is pushed across a surface. Through a process of negotiation, decisions are made by the group as to what might constitute appropriate experience for young children (such as pushing the car on different surfaces) and what might be inappropriate at this stage (the abstract concept of friction and use of this term). Although such insight may be obvious to the experienced practitioner, for inexperienced pre-service teachers we consider this to be discerning articulation of pedagogical content knowledge. While the SMLA approach clearly holds promise for stimulating PCK development within the institution-based setting, at present further research is needed in order to investigate the translation of this into classroom practice in the short and long term.

A number of interesting points emerge from the process that illustrate the SMLA potential in relation to developing subject matter pedagogy. Whilst the documenting and tracking of the dynamics of learning is difficult and requires a course philosophy based on enquiry, discourse and reflective practice, the case study shows that the SMLA enables students to scrutinise subject mater, recognise inherent challenges for learners and begin to consider implications for their own practice. In this sense they are able to identify and draw on knowledge bases that were both subject and pedagogic specific and identify the implications of this for how the content of the science curriculum could be represented in order to support pupils in understanding difficult counterintuitive ideas. Not surprisingly perhaps, the SMLA for Key Stage 1 proved more challenging than that focused on Key Stage 2. It is likely that the subject knowledge element in the former is not as easily identifiable and that the complex relation between progression in cognitive demand and associated pedagogic instruction are particularly elusive even for the most experienced of practitioners. PCK research has in the past been largely centred on secondary education; the professional challenge posed by the science education for young children remains to be explored.

The pre-service teachers in this study were capable of developing sophisticated insight into pedagogy outside of school-placed settings. This is shown in the identification of subtle nuances that underpin qualitative explanation. Not only does Sarah recognise that the inability to 'see' force acting and being transmitted within the simple arched bridge as a key difficulty in generating explanation, but she also acknowledges that the *realisation* of this is an important element of her own understanding. She also comes to recognise the difference between 'knowledge' and 'understanding 'and, furthermore, that although these terms are often presented simultaneously in curriculum documentation and guidelines they do not 'mean the same thing'.

The development of subject knowledge and related pedagogical knowledge plays an important role in the quest to produce a high-quality teaching force. Teacher preparation programmes worldwide contain such dimensions and the synthesis of the two determines the quality of translation in the classroom. Darling-Hammond and Cobb (1995) report that within APEC members, pedagogical content and subject area knowledge are important attributes of 'quality teachers' and, in a comprehensive review of research in the USA, Darling-Hammond (2000) concludes that teachers' knowledge, skills and preparation matter for children's achievement in school. In England, the standards for gaining qualified teacher status (TDA 2007) present clear expectations in this regard, requiring teachers to:

> [h]ave a secure knowledge and understanding of their subjects/curriculum areas and related pedagogy to enable them to teach effectively across the age and ability range for which they are trained. (TDA 2007:7)

There is an expectation that not only should beginning teachers possess sound knowledge of subject, but that they will also emerge from training with related pedagogical knowledge appropriate to delivering a 'personalised' curriculum that allows learners to achieve their potential.

Such aspirations imply that there exists an agreed pedagogy related to science subject matter to which the profession subscribes. Although there is undoubtedly wide-scale commitment to enquiry-based learning and the importance of learners' ideas in the process, much less is known of specific subject-related pedagogy. The

SMLA approach focuses attention on the cognitive demand of subject, the exploration of which contributes to developing PCK. Learning how to look critically at the curriculum nurtures transferable skills relevant to other areas of the curriculum containing equally challenging concepts. The *learning practice* provided by the SMLA experience has the potential to contribute to and augment the *teaching practice* that traditionally forms the foundation of teacher education.

7.6.2 Some Implications for the Role of Teacher Education Institutions

Finally, this research points to the unique role of higher education in providing opportunity to consider the link between personal learning and the explicit identification of pedagogic insight. In this process, individuals make explicit to themselves how they organise and synthesise the understanding of ideas in science (Traianou 2006). The articulation of this metacognition seems to be consequent on social interaction in which self-explanation derives from professional discourse where an individual, in explaining to others, clarifies their own understanding. This resonates with the notion that understanding derives from teaching. There is an interesting link here with the rationale underpinning why it is important for the teacher to have some awareness of children' ideas, not least because it provides insight into the relation between taxonomy in cognition and pedagogy.

Whilst teaching practice in school placements is a key element of teacher education programmes, learning practice offers potential to enhance teacher education through both complementing and extending such experiences. In the case of primary education, the opportunities to develop science PCK in school-based settings are limited and working environment does not necessarily facilitate this sort of knowledge development. The SMLA approach illustrates the possibility of a unique contribution to teacher learning that is dependent on the skill of the university tutor in terms of both knowledge of subject and pedagogy. It is simply not possible to provide opportunity for such in-depth learning consistently throughout a professional programme, but it may be beneficial for at least some exposure to such critical thinking once or twice during a professional course. This may be sufficient to alert intending teachers to the inherent difficulties of learning where ideas in science are abstract and counterintuitive and to the role of learners' ideas in the learning process and the critical importance of enquiry and discourse of learning as learners construct a qualitative understanding of subject through experience. Indeed the SMLA approach is predicated not on a deficit model of pre-service teachers' learning of science; rather, it is informed by what they know, understand and develop as a result of learning, thereby reinforcing ownership of the learning agenda; a strong motivational force that is particularly important in a subject such as science where learners typically lack confidence.

> I feel confident in my understanding of the subject/topics taught and now appreciate how important it is to put yourself in the place of a child to see how hard it is to learn difficult concepts. (Nasreen)

Gamache (2002) proposes that in order to enhance the learning of students in their education in general, they need to be helped to develop a different epistemological view, one in which they see themselves as creators of 'personal knowledge'. In science education this is likely to involve a radical reappraisal of not only science subject knowledge, but also epistemological views of what constitutes 'science' and 'science learning'. It follows that if students make such radical changes, both conceptually and epistemologically, then a reasonable starting point in the process would be for them to develop awareness of their own learning processes. Constructivist epistemology, with its focus on conceptual change approaches to teaching and learning science, usually requires of pre-service teachers a fundamental shift in personal beliefs about how learning takes place. In practice it has proved to be particularly difficult to influence beliefs about teaching and learning such that it results in changed classroom behaviour, and where progress has been made, this appears to be contingent on considerable periods of time and professional support for development (see for example, Tobin 1993; Summers and Kruger 1994; Wildy and Wallace 1995; Appleton and Asoko 1996). At present, the value of the SMLA experience in terms of its continued influence in planning and teaching science as the pre-service teacher enters the profession remains unknown and requires further research in order to establish its efficacy. We believe that the approach offers potential to allow space in a crowded teacher education curriculum to synthesise subject and pedagogy in a meaningful and useful way. The teaching decisions students make in real classrooms do not depend solely on subject matter. The ultimate expression of pedagogy will be influenced by a range of factors such as epistemological views, personal goals and beliefs and the constraints of the system in which they practice. It is also informed by their awareness of the knowledge, understanding, skills and attitudes of the children they work with and the teachers' understanding of the culture from which they are drawn. However, in generating pedagogical knowledge during initial teacher education there is possibility to influence beliefs about teaching and learning such that emerging beliefs are embedded in personal learning experience and that these are subsequently applied to teaching experience.

References

AAAS (1993). American Association for the Advancement of Science *Benchmarks for science literacy. Project 2061*. New York: Oxford University Press.

Abell, S. K. (2007). Research on science teacher knowledge. In S. K. Abell & N. G. Lederman (Eds.), *Handbook of research on science education* (pp. 1105–1150). Mahwah, NJ: Lawrence Erlbaum Associates.

Abell, S. K., & Lederman, N. G. (Eds.). (2007). *Handbook of research on science education*. Mahwah, NJ: Lawrence Erlbaum Associates.

Abell, S., Martini, M., & George, M. (2001). That's what scientists have to do: preservice elementary teachers' conceptions of the nature of science during a moon investigation. *International Journal of Science Education, 23*(11), 1095–1109.

Adams, P. E., & Krockover, G. H. (1997). Beginning science teacher cognition and its origins in the preservice secondary science program. *Journal of Research in Science Teaching, 34*, 633–653.

Adey, P., & Shayer, M. (1994). *Really raising standards*. London: Routledge.

Adey, P. S., Shayer, M., & Yates, C. (1995). *Thinking science: The curriculum materials of the CASE project*. Buckingham: Thomas Nelson & Son.

Akerson, V. L., Flick, L. B., & Lederman, N. G. (2000). The influence of primary children's ideas in science on teaching practice. *Journal of Research in Science Teaching, 37*(4), 363–385.

Ames, C. (1992). Classrooms: Goals, structures and student motivation in the middle grades. *Journal of Educational Psychology, 84*, 261–271.

Appleton, K. (2006). Science pedagogical content knowledge and elementary school teachers. In K. Appleton (Ed.), *Elementary science teacher education* (pp. 31–54). Mahwah, NJ: Lawrence Erlbaum Associates.

Appleton, K., & Asoko, H. (1996). A case study of a teacher's progress toward using a constructivist view of learning to inform teaching in elementary science. *Science Teacher Education, 80*(2), 165–180.

Atwood, V. A., & Atwood, R. K. (1995). Preservice elementary teachers' conceptions of what causes night and day. *School Science and Mathematics, 95*, 290–294.

Ausubel, D. P. (1968). *Educational psychology, a cognitive view*. New York: Holt, Rinehart and Winston.

Baddock, M., & Bucat, R. (2007). Effectiveness of a classroom chemistry demonstration using the cognitive conflict strategy. *International Journal of Science Education*, iFirst, online/08/000001-14. Retrieved March 2008.

Bailey, J. M., & Slater, T. F. (2003). A review of astronomy education research. *The Astronomy Education Review, 2*(2), 20–45.

Baird, J. R., & Mitchell, I. J. (eds). (1997). *Improving the quality of teaching and learning* (3rd ed.). Melbourne: PEEL.

Baldy, E. (2007). A new educational perspective for teaching gravity. *International Journal of Science Education, 29*(19), 1767–1788.

Baldy, E., & Aubert, F. (2005). Etude de l'apprentissage du phénomènephysique de la chute des corps en classe de troisième française, compte-rendu d'innovation. *Didaskalia, 27*, 109–132.

Ball, D. L., & McDairmid, G. W. (1996). The subject-matter preparation of teachers. In W. R. Houston (Ed.), *Handbook of research on teacher education* (pp. 437–449). New York: Macmillan.

Bar, V., Zinn, B., & Rubin, E. (1997). Children's ideas about action at a distance. *International Journal of Science Education, 19*(10), 1137–1157.

Barab, S. A., Barnett, M., Yamagata-Lynch, L., Squire, K., & Keating, T. (2002). Using activity theory to understand the systemic tensions characterizing a technology-rich introductory astronomy course. *Mind, Culture, and Activity, 9*(2), 76–107.

Barba, R., & Rubba, P. (1992). A comparison of pre-service and in-service earth and space science teachers' general mental abilities, content knowledge and problem solving skills. *Journal of Research in Science Teaching, 29*(10), 1021–1035.

Barnett, J., & Hodson, D. (2001). Pedagogical content knowledge: Toward a fuller understanding of what good teachers know. *Science Education, 85*(4), 426–453.

Barnett, M., MaKinster, J. G., & Hansen, J. A. (2001). *Exploring elementary students' learning of astronomy through model building*. Paper presented at the annual meeting of the American Educational Research Association, Seattle, WA. April 10–14.

Barnett, M., & Morran, J. (2002). Addressing children's alternative frameworks of the moon's phases and eclipses. *International Journal of Science Education, 24*(8), 859–879.

Baxter, J. (1989). Children's understanding of familiar astronomical events. *International Journal of Science Education, 11*, 502–513.

Beeth, M. E. (1998a). Teaching for conceptual change: Using status as a metacognitive tool. *Science Education, 82*(3), 343–356.

Beeth, M. E. (1998b). Facilitating conceptual change learning: The need for teachers to support metacognition. *Journal of Science Teacher Education, 9*(1), 49–61.

Beeth, M. E., & Hewson, P. W. (1999). Learning goals in an exemplary science teacher's practice: Cognitive and social factors in teaching for conceptual change. *Science Education, 83*, 738–760.

Bendall, S., Goldberg, F., & Galili, I. (1993). Prospective elementary teachers' prior knowledge about light. *Journal of Research in Science Teaching, 30*, 1169–1187.

Biggs, J. (1987). *Student approaches to learning and studying*. Melbourne: Australian Council for Educational Research.

Bisard, W., Aron, R., Francek, M., & Nelson, B. (1994). Assessing selected physical science and earth science misconceptions of middle school through university preservice teachers. *Journal of College Science Teaching, 24*, 38–42.

Black, P. J., & Harlen, W. (1993). How can we specify concepts for primary science ? In P. J. Black & A. M. Lucas (Eds.), *Children's informal ideas in science* (pp. 208–229). London: Routledge.

Blank, L. M. (2000). A metacognitive learning cycle: A better warranty for student understanding? *Science Education, 84*(4), 486–506.

Bloom, B. S. (1956). *Taxonomy of educational objectives, handbook 1: The cognitive domain.* New York: David McKay.

Borkowski, J. G. (1985). Sign of intelligence: Strategy generalisation and metacognition. In S. R. Yussen (Ed.), *The growth of reflection* (pp. 105–144). New York: Academic.

Botvin, G. J., & Murray, F. B. (1975). The efficacy of peer modeling and social conflict in the acquisition of conservation. *Child Development, 46*(3), 796–799.

BouJaoude, S. B. (1992). The relationship between students' learning strategies and the change in their misunderstandings during a high school chemistry course. *Research in Science Teaching, 29*(7), 689–699.

Brickhouse, N. W. (1990). Teachers' beliefs about the nature of science and their relationship to classroom practice. *Journal of Teacher Education, 41*(3), 53–62.

Brickhouse, N. W., & Bodner, G. M. (1992). The beginning science teacher: Classroom narratives of convictions and constraints. *Journal of Research in Science Teaching, 29*, 471–485.

Brown, A. L. (1987). Metacognition, executive control, self-regulation, and other mysterious mechanisms. In F. E. Weinert & R. H. Kluwe (Eds.), *Metacognition, motivation, and understanding* (pp. 65–116). Hillsdale, NJ: Lawrence Erlbaum Associates.

Brown, D. E. (1989). Students' concept of force: The importance of understanding Newton's third law. *Physics Education, 24*, 353–358.

Brown, T. (1997). *Mathematics education and language. Interpreting hermeneutics and post structuralism*. Dordrecht: Kluwer.

Brown, T. (2001). *Mathematics education and language*. Dordrecht: Kluwer.

Brown, D. E., & Clement, J. (1987). Misconceptions concerning Newton's law of action and reaction; the underestimated importance of the third law. In D. V. Novak (Ed.), *Proceedings of the second international seminar on misconceptions in science and mathematics* (Vol. 111, pp. 39–53). Ithaca, New York: Cornell University.

Browne, D. E. (1994). Facilitating conceptual change using analogies and explanatory models. *International Journal of Science Education, 16*(2), 201–214.

Bucat, R. (2004). Pedagogical content knowledge as a way forward: Applied research in chemistry education. *Chemistry Education: Research and Practice, 5*, 215–228.

Butler, D. (1998). Metacognition and learning disabilities. In B. Wong (Ed.), *Learning about learning disabilities* (pp. 277–307). San Diego, CA: Academic.

Calderhead, J. (1996). Teachers: Beliefs and knowledge. In D. C. Berliner & R. F. Chaffee (Eds.), *Handbook of educational psychology* (pp. 709–725). New York: Simon & Schuster Macmillan.

Carey, S. (1985). *Conceptual change in childhood*. Cambridge, MA: The MIT Press.

Carlsen, W. S. (2007). Language and science learning. In S. K. Abel & N. G. Lederman (Eds.), *Handbook of research on science education* (pp. 57–74). Mahwah, NJ: Lawrence Erlbaum Associates.

Cattle, J., & Howie, D. (2008). An evaluation of a school programme for the development of thinking skills through the CASE@KS1 approach. *International Journal of Science Education, 30*(2), 185–202.

CCCEA (2007). Council for Curriculum Examination and Assessment. *Nothern Ireland curriculum*. Council for Curriculum Examination and Assessment. Retrieved June 2009, from http://www.nicurriculum.org.uk

Chan, C., Burtis, J., & Bereiter, C. (1997). Knowledge building as a mediator of conceptual change. *Cognition and Instruction, 15*, 1–40.

Chi, M. T. H. (1992). Conceptual change in and across ontological categories: Examples from learning and discovery in science. In R. Giere (Ed.), *Cognitive models of science* (pp. 129–160). Minneapolis, MN: University of Minnesota Press.

Chi, M. T. H., & Roscoe, R. D. (2002). The processes and challenges of conceptual change. In M. Limòn and L. Mason (Eds.), *Reconsidering conceptual change. Issues in theory and practice* (pp. 3–28). Dordrecht: Kluwer.

Chi, M. T. H., Slotta, J. D., & De Leeuw, N. (1994). From things to process: A theory of conceptual change for learning science concepts. *Learning and Instruction, 4*, 27–43.

Chin, C. (2006). Classroom interaction in science: Teacher questioning and feedback to students' responses. *International Journal of Science Education, 28*(11), 1315–1346.

Chinn, C. A., & Brewer, W. F. (1993). The role of anomalous data in knowledge acquisition: A theoretical framework and implications for science education. *Review of Educational Research, 63*, 1–49.

Chinn, C. A., & Brewer, W. F. (1998). An empirical test of a taxonomy of responses to anomalous data in science. *Journal of Research in Science Teaching, 35*, 623–654.

Chin, C., & Brown, D. E. (2000). Learning in science: a comparison of deep and surface approaches. *Journal of Research in Science Teaching, 37*, 109–138.

Cin, M. (2007). Alternative views of the solar system among Turkish students. *International Review of Education, 53*(1), 39–53.

Clement, J. (1982). Students' preconceptions in introductory mechanics. *American Journal of Physics, 50*(1), 66–71.

Clement, J. (1993). Using bridging analogies and anchoring intuitions to deal with students' preconceptions in physics. *Journal of Research in Science Teaching, 30*(10), 1241–1257.

Clement, J. (2000). Model based learning as a key research area for science education. *International Journal of Science Education, 22*(9), 1041–1053.

Cochran-Smith, M., & Lytle, S. L. (1999). Relationships of knowledge and practice: teachers learning in communities. *Review of Research in Education, 24*, 249–305.

Connelly, F. M., & Clandinin, J. (1990). Stories of experience and narrative inquiry. *Educational Researcher, 19*(5), 2–14.

Cosgrove, M. (1995). A study in science-in-the-making as students generate an analogy for electricity. *International Journal of Science Education, 17*(3), 295–310.

Coward, R., & Ellis, J. (1977). *Language and materialism.* London: Routledge.

CSMEE (1996). Center for Science, Math and Engineering. *National Science Education Standards.* Retrieved June 2009, from httpp://www.nap.edu/openbook.php?record_id=4962.

Cutler, J. (2000). *Literary theory. A very short introduction.* Oxford: Oxford University Press.

Darling-Hammond, L. (2000). Teacher quality and student achievement: A review of state policy evidence. *Education Policy Analysis Archives*, 8(1). Retrieved accessed June 2009, from http://epaa.asu.edu/epaa/v8n1/.

Darling-Hammond, L., & Cobb, V. (1995). (Eds.). *Teacher preparation and professional development in APEC members: A comparative study.* Washington, DC: U.S. Department of Education. (ERIC Document Reproduction Service No. ED383683).

Davidowitz, B., & Rollnick, M. (2003). Enabling metacognition in the laboratory: A case study of four second year university chemistry students. *Research in Science Education, 33*, 43–69.

Day, C. (1993). Reflection: A necessary but not sufficient condition for teacher development. *British Educational Research Journal, 19*(1), 83–93.

De Jong, O., Van Driel, J. H., & Verloop, P. (2005). Preservice teachers' pedagogical content knowledge of using particle models in teaching chemistry. *International Journal of Science Education, 42*(8), 947–964.

Deci, E. L., Koestner, R., & Ryan, R. M. (2001). Extrinsic rewards and intrinsic motivation in education: Reconsidered once again. *Review of Educational Research, 71*, 1–27.

Deemer, S. A. (2004). Using achievement goal theory to translate psychological principles into practice in the secondary classroom. *American Secondary Education, 32*(3), 4–16.

Dekkers, P. J. J. M., & Thijs, G. D. (1998). Making productive use of students' initial conceptions in developing the concept of force. *Science Education, 82*, 31–51.

Demastes, S. S., Good, R. G., & Peebles, P. (1996). Patterns of conceptual change in evolution. *Journal of Research in Science Teaching, 33*, 407–431.

DfEE/QCA (1998). Department for Education and Employment/Qualifications and Curriculum Authority. *A scheme of work for key stages 1 and 2. Science.* London: Qualifications and Curriculum Authority.

DfEE/QCA (1999). Department for Education and Employment/Qualifications and Curriculum Authority. *The national curriculum for England. Science.* London: Department for Education and Employment and Qualifications and Curriculum Authority.

Dillon, C. (1994). Qualitative reasoning about physical systems - an overview. *Studies in Science Education, 23*, 39–57.

Disessa, A. A. (2002). Why 'conceptual ecology' is a good idea. In M. Limòn, & L. Mason (Eds.), *Reconsidering conceptual change. Issues in theory and practice* (pp. 29–60). Dordrecht: Kluwer.

DiSessa, A. A., & Sherrin, B. L. (1998). What changes in conceptual change? *International Journal of Science Education, 20*(10), 1155–1191.

Donaldson, M. (1978). *Children's minds.* London: Fontana.

Dreistadt, R. (1968). An analysis of the use of analogies and metaphors in science. *Journal of Psychology, 68*, 97–116.

Dreyfus, A., Jungwirth, E., & Eliovitch, R. (1990). Applying the 'cognitive conflict' strategy for conceptual change: Some implications, difficulties and problems. *Science Education, 74*, 555–569.

Driver, R. (1989a). Students' conceptions and the learning of science. *International Journal of Science Education, 11*, 481–490.

Driver, R. (1989b). Changing conceptions. In P. Adey (Ed.), *Adolescent development and school practice* (pp. 79–103). London: Falmer.

Driver, R. (1994). Children's ideas about physical processes. Electricity. In R. Driver, A. Squires, P. Rushworth, & V. Wood-Robinson (Eds.), *Making sense of secondary science* (pp. 117–125). London: Routledge.

Driver, R., & Easley, J. (1978). Pupils and paradigms: A review of literature related to concept development in adolescent science students. *Studies in Science Education, 5*, 61–84.

Driver, R., Squires, A., Rushworth, P., & Wood-Robinson, V. (eds). (1994). *Making sense of secondary science*. London: Routledge.

Duit, R. (2008). *Bibliography STCSE: Students' and teachers' conceptions and science education.* Kiel, Germany: IPN-Leibniz Institute for Science Education. Retrieved June 2008, from www.ipn.uni-kiel.de/aktuell/stcse/stcse.html.

Duit, R., & Treagust, D. F. (2003). Conceptual change: A powerful framework for improving science teaching and learning. *International Journal of Science Education, 25*(6), 671–688.

Dykstra, D. I. (1992). Studying conceptual change: Constructing new understandings. In: *Research in physics learning: Theoretical issues and empirical studies. Proceedings of an international workshop held at the University of Bremen, March 4–8 (1991).* Keil: Institut für die Padagogik der Naturwissenschaften an der Universitat.

Eagleton, T. (1997). *Literary theory. An introduction*. Oxford: Blackwell.

Easley, J. (1990). Stressing dialogic skill. In E. Duckworth, J. Easley & D. Hawkins (Eds.), *Science Education: A minds-on-approach for the elementary years* (pp. 61–93). Mahwah, NJ: Lawrence Erlbaum Associates.

Eger, M. (1992a). Hermeneutics and science education: An introduction. *Science and Education, 1*, 337–348.

Eger, M. (1992b). Hermeneutics as an approach to science: Part 1. *Science and Education*, 2, 1–29.

Eger, M. (1993). Hermeneutics as an approach to science: Part 2. *Science and Education*, 2(4), 303–328.

Eger, M. (2006). *Science, understanding and justice: The philosophical essays of Martin Eger.* In A. Shimony (Ed.). Chicago, IL: Open Court

Elizabeth, L. L., & Galloway, D. (1996). Conceptual links between cognitive acceleration through science education and motivational style: A critique of Adey and Shayer. *International Journal of Science Education, 18*, 35–49.

Everson, H. T., & Tobias, S. (1998). The ability to estimate knowledge and performance in college: A metacognitive analysis. *Instructional Science, 26*, 65–79.

Feher, E., & Rice, K. (1988). Shadows and anti-images: Children's conceptions of light and vision. II. *Science Education, 72*(5), 637–649.

Fensham, P. J. (2001). Science content as problematic – issues for research. In H. Behrendt, H. Dahncke, R. Duit, W. Graber, M. Komorek, A. Kross, et al. (Eds.), *Research in science education – past, present and future* (pp. 27–41). Dordrecht: Kluwer.

Fensham, P. J., Gunstone, R. F., & White, R. T. (1994). Science content and constructivist views of learning and teaching. In P. J. Fensham & R. T. White (Eds.), *The content of science: A constructivist approach to its teaching and learning* (pp. 1–8). London: Falmer.

Fenstermacher, G. D. (1994). The knower and the known: The nature of knowledge in research on teaching. In L. Darling-Hammond (Ed.), *Review of research in education (Vol 20)* (pp. 3–56). Washington, DC: American Educational Research Association.

Fernandez-Balboa, J. M., & Stiehl, J. (1995). The generic nature of pedagogical content knowledge among college professors. *Teaching and Teacher Education, 11*, 293–306.

Feynman, R. P. (1992). *The character of physical law*. London: Penguin.

Flavell, J. (1976). Metacognitive aspects of problem solving. In L. Resnick (Ed.), *The nature of intelligence* (pp. 231–235). Hillsdale, NJ: Lawrence Erlbaum Associates.

Flavell, J. (1979). Metacognition and cognitive monitoring: A new area of cognitive-developmental inquiry. *American Psychologist, 34*, 906–911.

Flavell, J. (1987). Speculations about the nature and development of metacognition. In F. Weinert & R. Kluwe (Eds.), *Metacognition, motivation and understanding* (pp. 20–29). Hillsdale, NJ: Lawrence Erlbaum Associates.

Fox, R. (2001). Constructivism examined. *Oxford Review of Education, 27*(1), 23–35.

Frederiksen, J. R., White, B. Y., & Guttwill, J. (1999). Dynamic mental models in learning science: the importance of constructing the derivational linkages among models. *Journal of Research in Science Teaching, 36*(7), 806–836.

Gadamer, H. G. (1993). *Truth and method.* London: Sheed & Ward.

Galili, I. (1996). Students' conceptual change in geometrical optics. *International Journal of Science Education, 18*, 847–868.

Galili, I. (2001). Weight versus gravitational force: Historical and educational perspectives. *International Journal of Science Education, 23*(10), 1073–1093.

Galili, I., & Bar, V. (1997). Children's operational knowledge about weight. *International Journal of Science Education, 19*, 317–490.

Galili, I., Bendall, S., & Goldberg, F. (1993). The effects of prior knowledge and instruction on understanding image formation. *Journal of Research in Science Teaching, 30*, 271–301.

Galili, I., & Hazon, A. (2000). Learners' knowledge in optics: Interpretation, structure and analysis. *International Journal of Science Education, 22*(1), 57–88.

Gallagher, S. (1992a). *Hermeneutics and education.* Albany, NY: State University of New York Press.

Gallagher, S. (1992b). Language and the imperfect consensus. In T. W. Busch & S. Gallagher (Eds.), *Merleau-Ponty, hermeneutics and postmodernism* (pp. 69–79). Albany, NY: State University of New York Press.

Gamache, P. (2002). University students as creators of personal knowledge: An alternative epistemological view. *Teaching in Higher Education, 7*(3), 277–294.

Gauld, C. F. (1986). Models, meters, and memory. *Research in Science Education, 16*, 49–54.

Gentner, D., & Gentner, D. R. (1983). Flowing waters or teeming crowds: mental models of electricity. In D. Gentner & A. Stevens (Eds.), *Mental models* (pp. 99–131). Mahwah, NJ: Lawrence Erlbaum Associates.

Georghiades, P. (2000). Beyond conceptual change learning in science education: focusing on transfer, durability and metacognition. *Educational Research, 42*, 119–139.

Georghiades, P. (2001). *Dimensions of meta-'Conceptual Change Learning' in science education: The role of metacognition in the durability and contextual use of pupils' conceptions.* Unpublished Ph.D. thesis, University of Surrey, Roehampton.

Georghiades, P. (2004a). Making pupils' conceptions of electricity more durable by means of situated metacognition. *International Journal of Science Education, 26*, 85–99.

Georghiades, P. (2004b). From the general to the situated: Three decades of metacognition. *International Journal of Science Education, 26*(3), 365–383.

Gess-Newsome, J. (1999). Pedagogical content knowledge: An introduction and orientation. In J. Gess-Newsome, & N. G. Lederman (Eds.), *Examining pedagogical content knowledge: the construct and its implications for science education* (pp. 51–94). Dordrecht, the Netherlands: Kluwer.

Gilbert, J. K. (2005). Visualization: A metacognitive skill in science and science education In J. K. Gilbert (Ed.), *Visualization in Science Education* (pp. 9–28). Netherlands: Springer.

Gilbert, J. K., Boulter, C., & Rutherford, M. (1988). Models in explanations: I Horses for courses? *International Journal of Science Education, 20*, 83–97.

P. Gilbert, J.K., Osborne, R.J., and Fensham, P. (1982) Children's science and its implications for teaching. *Science Education*, 66, 625–633.

Gilbert, J. K., & Reiner, M. (2000). Thought experiments in science education: Potential and current realization. *International Journal of Science Education, 22*(3), 265–283.

Gilbert, J. K., & Watts, M. (1983). Conceptions, misconceptions and alternative conceptions. *Studies in Science Education, 10*, 61–98.

Goldberg, F., & McDermott, L. C. (1986). Student difficulties in understanding image formation by a plane mirror. *The Physics Teacher, 24*, 472–480.

Goldberg, F., & McDermott, L. C. (1987). An investigation of student understanding of the real image formed by a converging lens or a concave mirror. *American Journal of Physics, 55*, 108–119.

Gooding, D. (1989). Magnetic curves' and the magnetic field. In D. Gooding, T. Pinch, & S. Schaffer, S. (Eds.), *The uses of experiment. Studies in the natural sciences* (pp. 183–224). Cambridge: Cambridge University Press.

Gordon, J. E. (1978). *Structures or why things don't fall down*. London: Penguin.

Gorsky, P., & Finegold, M. (1994). The role of anomaly and of cognitive dissonance in restructuring students' concepts of force. *Instructional Science, 22*, 75–90.

Gregory, B. (1988). *Inventing reality. Physics as language*. New York: Wiley.

Grossman, P. (1990). *The making of a teacher: Teacher knowledge and teacher education*. New York: Teachers College Press.

Guesne, E. (1985). Light. In R. Driver, E. Guesne & A. Tiberghien (Eds.), *Children's ideas in science* (pp. 10–32). Milton Keynes: Open University Press.

Gunstone, R. F. (1991). Reconstructing theory from practical experience. In B. E. Woolnough (Ed.), *Practical science* (pp. 66–77). Milton Keynes: Open University Press.

Gunstone, R. F., Slattery, M., Baird, J. R., & Northfield, J. R. (1993). A case study exploration of development in preservice science teachers. *Science Education, 77*(1), 47–73.

Hacker, D. (1998). Definitions and empirical foundations. In D. J. Hacker, J. Dunlossy & A. C. Graesser (Eds.), *Metacognition in educational theory and practice* (pp. 1–24). Mahwah, NJ: Lawrence Erlbaum Associates.

Hacker, D. J., Dunlossy, J., & Graesser, A. C. (eds). (1998). *Metacognition in educational theory and practice*. Mahwah, NJ: Lawrence Erlbaum Associates.

Halim, L., & Meerah, S. M. (2002). Science trainee teachers' pedagogical content knowledge and its influence on physics teaching. *Research in Science and Technological Education, 20*, 215–225.

Harland, R. (1994). *Superstructuralism*. London: Routledge.

Harlen, W. (2000). *The teaching of science in primary schools* (3rd ed.). London: David Fulton.

Harlen, W., & Holroyd, C. (1997). Primary teachers' understanding of the concepts of science: Impact on confidence and teaching. *International Journal of Science Education, 19*(1), 93–105.

Harrison, A. G., & Treagust, D. F. (2000). Conceptual change – a powerful framework for improving science teaching and learning. *International Journal of Science Education, 25*(6), 671–688.

Hashweh, M. Z. (1986). Toward an explanation of conceptual change. *European Journal of Science Education, 8*, 229–249.

Hewitt, P. G. (1989). *Conceptual physics*. San Francisco, CA: Harper Collins.

Hewson, P. W. (1981). A conceptual change approach to learning science. *European Journal of Science Education, 3*, 383–396.

Hewson, P. W. (1982). A case study of conceptual change in special relativity: The influence of prior knowledge in learning. *European Journal of Science Education, 4*, 61–78.

Hewson, P. W. (1996). Teaching for conceptual change. In D. F. Treagust & J. B. Fraser (Eds.), *Improving teaching and learning in science and mathematics* (pp. 131–140). New York: Teachers College Press.

Hewson, P. W., & Hewson, M. G. A'. B. (1984). The role of conceptual conflict in conceptual change and the design of science instruction. *Instructional Science, 13*, 1–13.

Hewson, P. W., & Hewson, M. G. A'. B. (1988). An appropriate conception of teaching science: A view from studies of learning. *Science Education, 75*(5), 597–614.

Hewson, P. W., & Hewson, M. G. A'B. (1992). The status of students' conceptions. In R. Duit, F. Goldberg, & H. Niedderer (Eds.), *Research in physics learning: Theoretical issues and empirical studies*. Proceedings of an international workshop (pp. 59–73). Kiel, Germany: IPN-Leibniz Institute for Science Education.

Heywood, D. (1999). *Interpretation and meaning in science education: Hermeneutic perspectives on language in learning and teaching science.* Ph.D. thesis, Manchester Metropolitan University, Manchester.

Heywood, D. (2002). The place of analogies in science education. *Cambridge Journal of Education, 32*(2), 233–248.

Heywood, D. (2005). Primary trainee teachers' learning and teaching about light: some pedagogical implications for initial teacher training. *International Journal of Science Education, 27*(12), 1447–1475.

Heywood, D. (2007). *Problematising science subject matter knowledge as a legitimate enterprise in primary teacher education.* Paper presented at the annual conference of the British Educational Research Association, London: Institute of Education, University of London, September 5–8.

Heywood, D., & Parker, J. (1997). Confronting the analogy: Primary teachers exploring the usefulness of analogies in the teaching and learning of electricity. *International Journal of Science Education, 19*, 869–885.

Heywood, D., & Parker, J. (2001). Describing the cognitive landscape in learning and teaching about forces. *International Journal of Science Education, 23*(11), 1177–1199.

Heywood, D., & Stronach, I. (2005). Philosophy and hermeneutics. In B. Somekh & C. Lewin (Eds.), *Research methods in the social sciences* (pp. 114–120). London: Sage.

Hidi, S., & Harackiewicz, J. M. (2000). Motivating the academically unmotivated: A critical issue for the 21st century. *Review of Educational Research, 70*, 151–179.

Hodson, D., & Hodson, J. (1998). From constructivism to social constructivism: A Vygotskian perspective on teaching and learning science. *School Science Review, 79*(2), 33–41.

Hofer, B. K., & Pintrich, P. R. (1997). The development of epistemological theories: Beliefs about knowledge and knowing and their relation to learning. *Review of Educational Research, 67*(1), 88–140.

Howard, B. C., & McGee, S. (2001). *The influence of metacognitive self-regulation and ability levels on problem-solving.* Paper presented at the annual conference of the American Educational Research Association. Seattle, WA, April 10–14.

Jenkins, E. W. (2004). Science Education: Research, practice and policy. In E. Scanlon, P. Murphy, J. Thomas & E. Whitelegg (Eds.), *Reconsidering science learning.* London: Routledge Falmer.

Jensen, M. S., & Finley, F. N. (1995). Teaching evolution using historical arguments in a conceptual change strategy. *Science Education, 79*(2), 147–166.

Johnson, P. (1998). Children's understanding of changes of state involving the gas state, part 2: Evaporation and condensation below boiling point. *International Journal of Science Education, 6*, 695–709.

Johnstone, A. H. (1997). Chemistry teaching – science or alchemy? *Journal of Chemical Education, 74*, 262–268.

Jones, E. M. (1999). *Critical pedagogy: an impossible task?* Unpublished Ph.D. thesis, Manchester Metropolitan University, Manchester.

Jung, W. (1993). Hilft die Entwicklungspsychologie dem Physikdidaktiker [Does developmental psychology help the physics educator?]. In R. Duit & W. Graber (Eds.), *Kognitive Entwicklung und naturwissenschaftlicher Unterricht* (pp. 86–107). Kiel, Germany: IPN – Leibniz Institute for Science Education.

Kang, S., Scarmann, L. C., & Noh, T. (2005). Re examining the role of cognitive conflict in science concept learning. *Research in Science Education, 34*(1), 71–96.

Kaplan, A., & Maehr, M. L. (1999). Enhancing the motivation of African American students: An achievement goal theory perspective. *The Journal of Negro Education, 68*, 23–42.

Keating, T., Barnett, M., Barab, S., & Hay, K. (2002). The Virtual Solar System project: Developing conceptual understanding of astronomical concepts through building three-dimensional computational models. *Journal of Science Education and Technology, 11*(2), 261–275.

Keil, F. C. (1979). *Semantic and conceptual development: An ontological perspective.* Cambridge, MA: Harvard University Press.

Kelly, G. J., & Chen, C. (1998). Students' reasoning about electricity: Combining performance assessments with argumentation analysis. *International Journal of Science Education, 20*(7), 849–871.

Kipnis, M., & Hofstein, A. (2007). The inquiry laboratory as a source for development of metacognitive skills. *International Journal of Science and Mathematics Education,* Online 1573–1774. National Science Council, Taiwan. Retrieved May 2008.

Klein, C. (1982). Children's concepts of the Earth and Sun: A cross-cultural study. *Science Education, 11*, 481–490.

Koch, A. (2001). Training in metacognition and comprehension of physics texts. *Science Education, 85*(6), 758–768.

Kruger, C., Summers, M. K., & Palacio, D. J. (1990). A survey of primary school teachers' conceptions of force and motion. *Educational Research, 32*(2), 83–94.

Kuhn, T. S. (1957). *The Copernican revolution.* Cambridge, MA: Harvard University Press.

Kuhn, T. S. (1970). *The structure of scientific revolutions* (2nd ed.). Chicago, IL: University of Chicago Press.

Lakatos, I. (1972). Falsification and the methodology of scientific research programs. In I. Lakatos & A. Musgrave (Eds.), *Criticism and the growth of knowledge.* London: Cambridge University Press.

Langley, D., Ronen, M., & Elyon, B. (1997). Light propagation and visual patterns: Pre instruction learners' conceptions. *Journal of Research in Science Teaching, 34*, 399–424.

Leach, J., & Scott, P. (2003). Individual and sociocultural views of learning in science education. *Science and Education, 12*(1), 91–113.

Lederman, N. G. (1999). Teachers' understanding of the nature of science and classroom practice: Factors that facilitate or impede the relationship. *Journal of Research in Science Teaching, 36*(8), 916–929.

Lederman, N. G., Gess-Newsome, J., & Latz, M. S. (1994). The nature and development of preservice science teachers' conceptions of subject matter and pedagogy. *Journal of Research in Science Teaching, 31*(2), 129–146.

Lee, G., Kwon, J., Park, S., Kim, J., Kwon, H., & Park, H. (2003). Development of an instrument for measuring cognitive conflict in secondary-level science class. *Journal of Research in Science Teaching, 40*(6), 585–603.

Lee, Y., & Taw, N. (2001). Explorations in promoting conceptual change in electrical concepts via ontological category shift. *International Journal of Science Education, 23*(2), 111–149.

Lehrer, R., & Schauble, L. (2006). Cultivating model-based reasoning in science education. In R. K. Sawyer (Ed.), *The Cambridge handbook of the learning sciences* (pp. 371–388). Cambridge: Cambridge University Press.

Lepper, M. R., & Hodell, M. (1989). Intrinsic motivation in the classroom. In C. Ames & R. Ames (Eds.), *Research on motivation in education* (Vol. 3, pp. 73–105). San Diego, CA: Academic.

Lightman, A., & Sadler, P. (1993). Teacher predictions versus actual student gains. *The Physics Teacher, 31*, 162–167.

Limón, M. (2001). On the cognitive conflict as an instructional strategy for conceptual change: A critical appraisal. *Learning and Instruction, 11*, 357–380.

Limón, M., & Carretero, M. (1997). Conceptual change and anomalous data: A case study in the domain of natural sciences. *Educational Journal of Psychology of Education, 12*(2), 213–230.

Limón, M., & Mason, L., (Eds.), (2002). *Reconsidering conceptual change: Issues in theory and practice* (pp. 61–76). Dordrecht, the Netherlands: Kluwer.

Livingston, J. A. (2003). Metacognition: An Overview. Retrieved April 2008, from eric.ed.gov. ERIC #: ED474273.

Loughran, J. J. (2002). Effective reflective practice: In search of meaning in learning about teaching. *Journal of Teacher Education, 53*(1), 33–43.

Loughran, J. J., Mitchell, I. J., & Mitchell, J. (eds). (2002). *Learning from teacher research.* New York: Teachers College Press.

Lunn, S. (2002). 'What We Think We Can Safely Say…': primary teachers' views of the nature of science. *British Educational Research Journal, 28*(5), 649–672.

Magnusson, S., Krajeik, J., & Borko, H. (1999). Nature, sources and development of pedagogical content knowledge. In J. Gess-Newsome, & N. G. Lederman (Eds.), *Examining pedagogical content knowledge: The construct and its implications for science education* (pp. 95–132). Dordrecht, the Netherlands: Kluwer.

Mali, G. B., & Howe, A. (1979). Development of Earth and gravity concepts among Nepali children. *Science Education, 63*(5), 685–691.

Marks, R. (1990). Pedagogical content knowledge: From a mathematical case to a modified conception. *Journal of Teacher Education, 41*(3), 3–11.

Marton, F. (1983). Beyond individual differences. *Educational Psychology, 3*, 289–303.

Marton, F., & Säljö, R. (1976). On qualitative differences in learning 1: outcome and process. *British Journal of Educational Psychology, 46*, 4–11.

Mason, L. (2000). Role of anomalous data and epistemological beliefs in middle school students' theory change about two controversial topics. *European Journal of Psychology of Education, 15*(2), 329–346.

Mathewson, J. H. (1999). Visual-spatial thinking: An aspect of science overlooked by educators. *Science Education, 83*(1), 33–54.

Matthews, P. S. C. (2000). Learning science: Some insights from cognitive science. *Science and Education, 9*, 507–535.

McCloskey, M. (1983). Intuitive physics. *Scientific American, 248*, 122–130.

McCormick, C. B. (2003). Metacognition and learning. In W. M. Reynolds, G. E. Miller & I. B. Weiner (Eds.), *Current perspectives in educational psychology volume 7, educational psychology* (pp. 79–97). New York: Wiley.

McGuinness, C. (1999). *From thinking skills to thinking classrooms: A review and evaluation of approaches for developing pupils' thinking*. Nottingham: DfEE.

McMillan, J. H., & Schumacher, S. (2001). *Research in education*. New York: Addison Wesley Longman.

McNamara, O. (1995). *The construction of knowledge in mathematics education*. Unpublished Ph.D. thesis, Manchester Metropolitan University, Manchester.

Mercer, N., Dawes, L., Wegerif, R., & Sams, C. (2004). Reasoning as a scientist: ways of helping children to use language to learn science. *British Educational Research Journal, 30*(3), 359–377.

Metz, K. (1991). Development of explanation: Incremental and fundamental changes in children's physics knowledge. *Journal of Research in Science Teaching, 28*, 785–798.

Mikropoulos, T. A. (2006). Presence: A unique characteristic in educational virtual environments. *Virtual Reality, 10*(3–4), 197–206.

Minstrell, J. (1992). Facets of students' knowledge and relevant instruction. In R. Duit, F. Goldberg, & H. Neidderer, (Eds.), *Research in physics learning: Theoretical issues and empirical studies* (pp. 110–128). Kiel: Kiel, Germany: IPN-Leibniz Institute for Science Education.

MoE, PRC (2001). Ministry of Education, People's Republic of China. *Guidelines for the curriculum reform of basic education*. Retrieved June 2009, from http://www.moe.edu.cn/eduas/ web site18/en/basic_b.htm.

MoE, NZ (2007). Ministry of Education, New Zealand. *The New Zealand curriculum learning media limited*. Wellington, NZ: Learning Media.

Mortimer, E. F., & Scott, P. H. (2003). *Meaning making in secondary science classrooms*. Buckingham: Open University Press.

Movshovitz-Hadar, N., & Hadass, R. (1990). Preservice education of math teachers using paradoxes. *Educational Studies in Mathematics, 21*, 265–287.

Mulholland, J., & Ginns, I. (2008). College MOON Project Australia: Preservice teachers learning about the Moon's phases. *Research in Science Education, 38*, 385–399.

Mulholland, J., & Wallace, J. (2005). Growing the tree of teacher knowledge: Ten years of learning to teach elementary science. *Journal of Research in Science Teaching, 42*(7), 767–790.

Munby, H., Russell, T., & Martin, A. K. (2001). Teachers' knowledge and how it develops. In V. Richardson (Ed.), *Handbook of research on teaching* (4th ed., pp. 877–904). Washington, DC: American Educational Research Association.

National Research Council (NRC). (1996). *National science education standards*. Washington, DC: National Academy Press.

Niaz, M. (1995). Cognitive conflict as a teaching strategy in solving chemistry problems: A dialectic – constructivist perspective. *Journal of Research in Science Teaching, 32*, 959–970.

Nussbaum, J. (1985). The Earth as a cosmic body. In R. Driver, E. Guesne & A. Tiberghien (Eds.), *Children's ideas in science* (pp. 170–192). Milton Keynes: Open University Press.

Nussbaum, J., & Novak, J. D. (1976). An assessment of children's concepts of the earth utilizing structured interviews. *Science Education, 60*(4), 535–550.

Ofsted (2007). Office for Standards in Education. *Success in science.* Retrieved June 2008, from http://www.ofsted.gov.uk/education/119122.

Ofsted (2008). Office for Standards in Education. *Outstanding School Report 070195.* Retrieved June 2008, from http://www.ofsted.gov.uk/portal/site/Internet/menuitem.eace3f09a603f6d9c3172a8a08c08a0c/.

Ogborn, J., Kress, G., Martins, I., & McGillicuddy, K. (1996). *Explaining science in the classroom.* Buckingham: Open University Press.

Osborne, J., Black, P., Smith, M., & Meadows, J. (1991). *Science processes and concept exploration project. Research report. Electricity.* Liverpool: Liverpool University Press.

Osborne, R., & Freyberg, P. (Eds.). (1985). *Learning in science: The implications of 'children's science'.* London: Heinemann.

Palmer, D. (2001). Students' alternative conceptions and scientifically acceptable conceptions about gravity. *International Journal of Science Education, 23*(7), 691–706.

Palmer, D. (2005). A motivational view of constructivist-informed teaching. *International Journal of Science Education, 27*(15), 1853–1881.

Paris, S. G., & Winograd, P. (1990). How metacognition can promote academic learning and instruction. In B. F. Jones, & L. Idol (Eds.), *Dimensions of thinking and cognitive instruction* (pp. 1551). Hillsdale, NJ: Lawrence Erlbaum Associates.

Park, D. (1997). *The fire within the eye. An historical essay on the nature and meaning of light.* Princeton, NJ: Princeton University Press.

Park, J. (2006). Modelling analysis of students' processes of generating scientific explanatory hypotheses. *International Journal of Science Education, 28*(5), 469–489.

Park, J., Ikgyun, K., Myunghwan, K., & Lee, M. (2001). Analysis of students' processes of confirmation and falsification of their prior ideas about electrostatics. *International Journal of Science Education, 23*(12), 1219–1236.

Park, J., & Pak, S. (1997). Students' responses to experimental evidence based on perceptions of causality and availability of evidence. *Journal of Research in Science Teaching, 34*(1), 57–67.

Parker, J. (2006). Exploring the impact of varying degrees of cognitive conflict in the generation of both subject and pedagogical knowledge as primary trainee teachers learn about shadow formation. *International Journal of Science Eduaction, 28*(13), 1545–1577.

Parker, J., & Heywood, D. (1998). The earth and beyond: Developing primary teachers' understanding of basic astronomical events. *International Journal of Science Education, 22*(5), 503–520.

Parker, J., & Heywood, D. (2000). Exploring the relationship between subject knowledge and pedagogic content knowledge in primary teachers' learning about force. *International Journal of Science Education, 22*, 89–111.

Parker, J., & Spink, E. (1997). Becoming science teachers: An evaluation of the initial stages of primary teacher training. *Assessment and Evaluation in Higher Education, 22*(1), 17–31.

Phillips, D. C. (2000). *Constructivism in education: opinions and second opinions on controversial issues.* Chicago, IL: Chicago University Press.

Piaget, J. (1914). Bergson et Sabatier. *Revue chrétienne (Paris), 61*(4), 192–200.

Piaget, J. (1964). *Six psychological studies.* New York: Vintage.

Piaget, J. (1985). *The equilibrium of cognitive structures.* Chicago, IL: University of Chicago Press.

Piaget, J., & Inhelder, B. (1956). *The child's conception of space.* London: Routledge.

Pintrich, P. R. (2000). Multiple goals, multiple pathways: The role of goal orientation in learning and achievement. *Journal of Educational Psychology, 92*, 544–555.

Pintrich, P. R., Marx, R. W., & Boyle, R. A. (1993). Beyond cold conceptual change: The role of motivational beliefs and classroom contextual factors in the process of conceptual change. *Review of Educational Research, 63*, 167–199.

Pintrich, P., Wolters, C., & Baxter, G. (2000). Assessing metacognition and self regulated learning. In G. Schraw & J. Impara (Eds.), *Issues in the measurement of metacognition* (pp. 43–146). Lincoln, NE: Buros Institute of Mental Measurements.

Plowden, B. (1967). *The Plowden Report: Children and their primary schools*. London: HMSO.

Popper, K. R. (1963). *Conjectures and refutations*. London: Routledge.

Posner, G. J., Strike, K. A., Hewson, P. W., & Gertzog, W. A. (1982a). Accommodation of a scientific conception: Toward a theory of conceptual change. *Science Education, 66*(2), 211–227.

Poulson, L. (2001). Paradigm lost? Subject knowledge, primary teachers and education policy. *British Journal of Educational Studies, 49*(1), 40–55.

Reiner, M. (1997). The context of thought experiments in physics learning. *Interchange, 37*(1–2), 97–113.

Reiner, M., & Gilbert, J. (2000). Epistemological resources for thought experimentation in science learning. *International Journal of Science Education, 22*(5), 489–506.

Reiner, M., & Gilbert, J. (2004). The symbiotic roles of empirical experimentation and thought experimentation in the learning of physics. *International Journal of Science Education, 26*(15), 1819–1834.

Reynolds, M. C. (1989). *Knowledge base for the beginning teacher*. Oxford: Pergamon.

Rice, K., & Feher, E. (1987). Pinholes and images: Children's conceptions of light and vision. *Science Education, 7*, 629–639.

Roth, W. M. (1996). Knowledge diffusion in a grade 4–5 classroom during a unit of civil engineering: An analysis of a classroom community in terms of its changing resources and practices. *Cognition and Instruction, 14*, 170–220.

Rowlands, M. A. (2006). What is buoyancy force? *Journal of Science Education, 7*(1), 39–42.

Russell, T., McGuigan, L., & Hughes, A. (1998). *Science processes and concept exploration (SPACE) research report: Forces*. Liverpool: Liverpool University Press.

Ryan, R. M., & Deci, E. L. (2000). Intrinsic and extrinsic motivations: Classic definitions and new directions. *Contemporary Educational Psychology, 25*, 54–67.

Samarapungavan, A., Vosniadou, S., & Brewer, W. F. (1996). Mental models of the Earth, Sun, and Moon: Indian children's cosmologies. *Cognitive Development, 11*(4), 491.

Sanders, L. R., Borko, H., & Lockard, J. D. (1993). Secondary science teachers' knowledge base when teaching out of their area of certification. *Journal of Research in Science Teaching, 31*, 723–736.

Saussure, F.de (1966). *A course in general linguistics*. New York: McGraw-Hill.

Schneps, M. P. (1989). *A private universe, Video*. San Francisco, CA: Astronomical Society of the Pacific.

Schön, D. A. (1987). *Educating the reflective practitioner*. San Francisco, CA: Jossey-Bass.

Schoultz, J., Säljö, R., & Wyndham, J. (2001). Heavenly talk: Discourse, artefacts and children's understanding of astronomy. *Human Development, 44*, 103–118.

Schraw, G. (1995). Measures of feeling-of-knowing accuracy: A new look at an old problem. *Applied Cognitive Psychology, 9*, 321–332.

Schraw, G. (1998). Promoting general metacognitive awareness. *Instructional Science, 26*, 113–125.

Schraw, G., & Moshman, D. (1995). Metacognitive theories. *Educational Psychology Review, 7*, 351–371.

Schwab, J. J. (1964). The structure of the disciplines. In G. Ford & L. Purgo (Eds.), *The structure of knowledge and the curriculum* (pp. 6–30). Chicago, IL: Rand McNally.

Schwab, J. J. (1978). Education and the structure of the disciplines. In I. Westbury & N. J. Wilkof (Eds.), *Science, curriculum and liberal education* (pp. 229–272). Chicago, IL: University of Chicago Press.

Scott, P., Asoko, H., & Leach, J. (2007). Student conceptions and conceptual learning in science. In S. K. Abell & N. G. Lederman (Eds.), *Handbook of research on science education* (pp. 31–56). Mahwah, NJ: Lawrence Erlbaum Associates.

Scott, P., Mortimer, E., & Aguiar, O. (2006). The tension between authoritative and dialogic discourse: a fundamental characteristic of meaning making interactions in high school science lessons. *Science Education, 90*, 605–631.

Sfard, A. (1998). On two metaphors of learning and the dangers of choosing just one. *Educational Researcher, 27*(2), 4–13.

Sharp, J. G. (1996). Children's astronomical beliefs: a preliminary study of year 6 children in south-west England. *International Journal of Science Education, 18*(6), 685–712.

Sharp, J. G., & Kuerbis, P. (2005). Children's ideas about the solar system and the chaos in learning science. *Science Education, 89*, 1–25.

Sharp, J. G., Peacock, G., Johnsey, R., Simon, S., & Smith, R. (2000). *Primary science teaching theory and practice*. Exeter: Learning Matters.

Shayer, M., & Adey, P. (2002). *Learning intelligence: Cognitive acceleration across the curriculum from 5 to 15 years*. Milton Keynes: Open University Press.

Shepardson, D. P., & Moje, E. B. (1999). The role of anomalous data in restructuring fourth graders' frameworks for understanding electric circuits. *International Journal of Science Education, 21*(1), 77–94.

Shipstone, M. (1984). A study of children's understanding of electricity in simple D.C. circuits. *European Journal of Science Education*, 6, 185–195.

Shulman, L. S. (1986). Those who understand: knowledge growth in teaching. *Educational Researcher, 15*(2), 4–14.

Shulman, L. S. (1987). Knowledge and teaching: foundations on the new reform. *Harvard Educational Review, 57*(1), 1–22.

Sinatra, G. M., & Pintrich, P. R. (eds). (2003). *Intentional conceptual change*. Mahwah, NJ: Lawrence Erlbaum.

Smith, D. C., & Neale, D. C. (1991). The construction of subject-matter knowledge in primary science teaching. In J. Brophy (Ed.), *Advances in research on teaching* (pp. 187–243). Greenwich: JAI.

Solomon, J. (1986). Children's explanations. *Oxford Review of Education, 12*(1), 41–51.

Solomon, J. (1994). The rise and fall of constructivism. *Studies in Science Education, 23*, 1–19.

Son, L., & Schwartz, B. (2002). The relation between metacognitive monitoring and control. In T. Perfect & B. Schwartz (Eds.), *Applied Metacognition* (pp. 15–38). Cambridge: Cambridge University Press.

Sorensen, R. A. (1992). *Thought experiments*. New York: Oxford University Press.

Spiliotopoulou-Papantoniou, V. (2007). Models of the universe: Children's experiences and evidence from the history of science. *Science and Education, 16*(7–8), 801–833.

Stahly, L. L., Krockover, G. H., & Shepardson, D. P. (1999). Third grade students' ideas about the lunar phases. *Journal of Research in Science Teaching, 36*(2), 159.

Stavy, R., & Berkowitz, B. (1980). Cognitive conflicts as a basis for teaching quantitative aspects of the concept of temperature. *Science Education, 64*, 679–692.

Stavy, R., & Tirosh, D. (2000). *How students (mis-)understand science and mathematics: Intuitive rules*. New York: Teachers College Press.

Sternberg, R. (1998). Metacognition, abilities and developing expertise: What makes an expert student? *Instructional Science, 26*, 127–140.

Strike, K. A. (1987). Toward a coherent constructivism. In J. Novak, (Ed.), *Proceedings of the 2. Int. Seminar Misconceptions and Educational Strategies in Science and Mathematics*, Vol. I. 9 (pp. 481–489). Ithaca, NY: Cornell University.

Strike, K. A., & Posner, G. J. (1982). Conceptual change and science teaching. *European Journal of Science Education, 4*, 231–240.

Strike, K. A., & Posner, J. G. (1985). A conceptual change view of learning and understanding. In L. West & R. Hamilton (Eds.), *Cognitive structure and conceptual change* (pp. 211–232). London: Academic.

Strike, K. A., & Posner, G. J. (1992). A revisionist theory of conceptual change. In R. Duschl & R. Hamilton (Eds.), *Philosophy of science, cognitive psychology and educational theory and practice* (pp. 147–176). Albany, NY: State University of New York Press.

Sturrock, J. (1993). *Structuralism*. London: Fontana.

Summers, M. (1994). Science in the primary school: the problem of teachers' curricular expertise. *The Curriculum Journal, 5*(2), 179–193.

Summers, M., & Kruger, C. (1994). A longitudinal study of a constructivist approach to improving primary science teachers' subject matter knowledge in science. *Teaching and Teacher Education, 10*, 499–519.

Summers, M., Kruger, C., & Mant, J. (1998). Teaching electricity effectively in the primary school: a case study. *International Journal of Science Education, 20*(2), 153–172.

Summers, M., & Mant, J. (1995). A survey of some primary school teachers' understanding of the Earth's place in the universe. *Educational Research, 37*(1), 3–19.

Tao, P., & Gunstone, R. F. (2000). The process of conceptual change in force and motion during computer supported physics instruction. *Journal of Research in Science and Technology, 36*(7), 859–882.

Targan, D. (1987). A study of conceptual change in the content domain of the lunar phases. *Proceedings of the Second International Seminar on Misconceptions and Educational Strategies in Science and Maths,* 2, 499–511. Ithaca, NY: Cornell University Press.

Tasker, R., & Osborne, R. (1985). Science teaching and science learning. In R. Osborne & P. Freyberg (Eds.), *Learning in science, the implications of children's science*. Auckland, NZ: Heinemann.

TDA (2007). Training and Development Agency for schools. *The revised standards for the recommendation for qualified teacher status (QTS).* Retrieved Feb. 2007, from http://www.tda.gov.uk/upload/resources/doc/draft_qts_standards.

Thorley, N. R., & Treagust, D. F. (1989). Conflict within dyadic interactions as a stimulant for conceptual change in Physics. *International Journal of Science Education, 9*(2), 203–216.

Tiberghein, A. (1985). Some features of children's ideas and their implications for teaching. In R. Driver, E. Guesne, & A. Tiberghein (Eds.), *Children's ideas in science* (pp. 1–9, 193–201). Buckingham: Open University Press.

Tobin, K. (ed). (1993). *The practice of constructivism in science education*. Washington, DC: AAAS.

Toulmin, S. (1972). *Human understanding Vol. 1: The collective use and evolution of concepts.* Princeton, NJ: University of Princeton Press.

Traianou, A. (2006). Teachers' adequacy of subject knowledge in primary science: Assessing constructivist perspectives from a sociocultural perspective. *International Journal of Science Education, 28*(8), 827–842.

Treagust, D. F. (2007). General instructional methods and strategies. In S. K. Abell & L. G. Lederman (Eds.), *Handbook of research on science education* (pp. 373–392). Mahwah, NJ: Lawrence Erlbaum Associates.

Trumper, R. (1997). Applying conceptual conflict strategies in the learning of the energy concept. *Research in Science and Technology Education, 15*, 5–18.

Trumper, R. (2001a). A cross-college age study of science and nonscience students' conceptions of basic astronomy concepts in preservice training for high-school teachers. *Research in Science and Technological Education, 10*(2), 189–195.

Trumper, R. (2001b). Assessing students' basic astronomy conceptions from junior high school through university. *Australian Science Teachers Journal, 41*, 21–31.

Trumper, R. (2003). The need for change in elementary school teacher training: A cross-college age study of future teachers' conceptions of basic astronomy concepts. *Teachers and Teacher Education, 19*(3), 309–323.

Trumper, R. (2006). Teaching future teachers basic astronomy concepts – seasonal changes-at a time of reform in science education. *Journal of Research in Science Teaching, 43*(9), 879–906.

Trumper, R., & Gorsky, P. (1996). A survey of biology students' conceptions of force in pre-service training for high school teachers. *Research in Science and Technological Education, 15*(2), 133–147.

Trumper, R., & Gorsky, P. (1997). A survey of biology students' conceptions of force in pre-service training for high school teachers. *Research in Science and Technological Education, 15*, 133–147.

Trundle, K. C., Atwood, R. K., & Christopher, J. E. (2002). Preservice elementary teachers' conceptions of moon phases before and after instruction. *Journal of Research in Science Teaching, 39*(7), 633.

Turner-Bisset, R. A. (1999). The knowledge bases of the expert teacher. *British Educational Research Journal, 25*(1), 39–55.

Turner-Bisset, R. A. (2001). *Expert teaching: Knowledge and teaching to lead the profession.* London: Fulton.

Van Driel, J. H., Verloop, N., & De Vos, W. (1998). Developing science teachers' pedagogical content knowledge. *Journal of Research in Science Teaching, 35*(6), 673–695.

Van Zee, E. H., & Roberts, D. (2001). Using pedagogical inquiries as a basis for learning to teach: Prospective teachers' perceptions of positive science learning experiences. *Science Education, 85*, 733–757.

Veenman, M. V. J., Prins, F. J., & Elshout, J. J. (2002). Initial inductive learning in a complex computer simulated environment: The role of metacognitive skills and intellectual ability. *Computers in Human Behavior, 18*, 327–341.

Viennot, L. (1979). Spontaneous reasoning in elementary mechanics. *European Journal of Science Education, 1*, 205–221.

Viennot, L., & Rainsom, S. (1999). Design and evaluation of a research based teaching sequence. The superposition of electrical fields. *International Journal of Science Education, 21*, 1–6.

Von Glasersfeld, E. (1992). A constructivist's view of learning and teaching. In R. Duit, F. Goldberg & H. Niedderer (Eds.), *Research in Physics Learning: Theoretical Issues and Empirical Studies* (pp. 29–39). University of Kiel: Institute for Science Education.

Vosnaidou, S. (1991). Designing curricula for conceptual restructuring: lessons from the study of knowledge acquisition in astronomy. *Curriculum Studies, 23*(3), 219–237.

Vosnaidou, S. (1994). Universal and culture-specific properties of children's mental models of the earth. In L. Hirschfeld & S. Gelman (Eds.), *Mapping the mind* (pp. 412–430). Cambridge: Cambridge University Press.

Vosnaidou, S. (2001). Conceptual change research and the teaching of science. In H. Behrendt, H. Dahncke, R. Duit, W. Graber, M. Komorek, A. Kross & P. Reiske (Eds.), *Research in science education – past, present, and future* (pp. 177–188). Dordrecht: Kluwer.

Vosniadou, S. (2002). On the nature of naive physics. In M. Limòn, & L. Mason (Eds.), *Reconsidering the processes of conceptual change* (pp. 61–76). Dordrecht, the Netherlands: Kluwer.

Vosniadou, S., & Brewer, W. F. (1992). Mental models of the earth: A study of conceptual change in childhood. *Cognitive Psychology, 24*, 535–585.

Vosniadou, S., & Ioannides, C. (1998). From conceptual development to science education: A psychological point of view. *International Journal of Science Education, 20*(10), 1213–1230.

Vosnaidou, S., Ionnides, C., Dimitrakopoulou, A., & Papademetriou, E. (2001). Designing learning environments to promote conceptual change in science. *Learning and Instruction, 11*(4–5), 381–419.

Vygotsky, L. S. (1978). *Mind in society: The development of higher psychological processes.* Cambridge, MA: Harvard University Press.

Walkerdine, V. (1988). *The mastery of reason*. London: Routledge.

Wang, M., Haertel, G., & Walberg, H. (1990). What influences learning? A content analysis of review literature. *Journal of Educational Research, 84*, 30–44.

Watts, D. M. (1983). A study of schoolchildren's alternative frameworks of the concept of force. *International Journal of Science Education, 5*(2), 217–230.

Wegerif, R. (2008). Dialogic or dialectic? The significance of ontological assumption in research on educational dialogue. *British Educational Research Journal, 34*(3), 347–361.

Weinert, F. E. (1987). Metacognition and motivation as determinants of effective learning and understanding. In F. E. Weinert & R. H. Kluwe (Eds.), *Metacognition, motivation, and understanding* (pp. 1–16). Hillsdale, NJ: Lawrence Erlbaum Associates.

Wells, G. (1999). *Dialogic inquiry. Toward a sociocultural practice and theory of education.* Cambridge: Cambridge University Press.

Wenden, A. (1999). Metacognitive knowledge and language learning. *Applied Linguistics, 19*(4), 515–537.

Wertsch, J. V. (1991). *Voices of the mind: A sociocultural approach to mediated action*. Cambridge, MA: Harvard University Press.

White, B., & Frederiksen, J. (2005). A theoretical framework and approach for fostering metacognitive. *Development Educational Psychologist, 40*(4), 211–223.

White, R. T., & Mitchell, I. J. (1994). Metacognition and the quality of learning. *Studies in Science Education, 23*, 21–37.

Wilbers, J., & Duit, R. (2001). On the micro-structure of analogical reasoning: The case of understanding chaotic systems. In H. Behrendt, H. Dahncke, R. Duit, W. Graber, M. Komorek, A. Kross, et al. (Eds.), *Research in science education – past, present, and future* (pp. 205–210). Dordrecht: Kluwer.

Wildy, H., & Wallace, J. (1995). Understanding teaching or teaching for understanding: Alternative frameworks for science classrooms. *Journal of Research in Science Teaching, 32*, 143–156.

Wilson, S. M., Shulman, L. S., & Richert, A. E. (1987). 150 Different ways of knowing: Representations of knowledge in teaching. In J. Calderhead (Ed.), *Exploring teacher' thinking* (pp. 104–124). London: Cassell.

Wiser, M., & Amin, T. G. (2001). 'Is heat hot?' Inducing conceptual change by integrating everyday and scientific perspectives on thermal phenomena. *Learning and Instruction, 11*, 331–355.

Wolpert, L. W. (1992). *The unnatural nature of science*. London: Faber & Faber.

Wong, D. E. (1993). Understanding the generative capacity as analogies as a tool for explanation. *Journal of Research in Science Teaching, 30*(10), 1259–72.

Woodruff, E., & Meyer, K. (1997). Explanations from intra- and inter-group discourse: Students building knowledge in the science classroom. *Research in Science Education, 27*, 25–39.

Wubbels, T. (1992). Taking account of student teachers' preconceptions. *Teaching and Teacher Education, 8*(2), 137–114.

Xiang, P., McBride, R., & Solomon, M. A. (2003). Motivational climates in ten teachers' elementary physical education classes: An achievement goal theory approach. *The Elementary School Journal, 104*, 71–92.

Yu, K. C. (2005). Digital full-domes: The future of virtual astronomy education planetarian. *Journal of the International Planetarium Society, 34*(3), 6–11.

Zimmerman, B. J., & Blom, D. E. (1983). Toward an empirical test of the role cognitive conflict in learning. *Developmental Review, 3*, 18–38.

Zohar, A., & Aharon-Kravetsky, S. (2005). Explaining the effects of cognitive conflict and direct teaching for students of different academic levels. *Journal of Research in Science Teaching, 42*(7), 829–855.

Author Index

Subject Index

FSC
www.fsc.org
MIX
Papier | Fördert gute Waldnutzung
FSC® C083411

Kolibri 360 GmbH
Ferdinand-Jühlke-Straße 7
99095 Erfurt, Deutschland
produktsicherheit@kolibri360.de